Essentials of Metallurgy

Essentials of Metallurgy

Edited by
Chloe Raines

Larsen & Keller
www.larsen-keller.com

Essentials of Metallurgy
Edited by Chloe Raines
ISBN: 978-1-63549-701-4 (Hardback)

© 2018 Larsen & Keller

☰ Larsen & Keller

Published by Larsen and Keller Education,
5 Penn Plaza,
19th Floor,
New York, NY 10001, USA

Cataloging-in-Publication Data

Essentials of metallurgy / edited by Chloe Raines.
    p. cm.
Includes bibliographical references and index.
ISBN 978-1-63549-701-4
1. Metallurgy. 2. Metals. I. Raines, Chloe.
TN665 .E87 2018
669--dc23

For more information regarding Larsen and Keller Education and its products, please visit the publisher's website www.larsen-keller.com

Table of Contents

Preface

The study and examination of the chemical and physical properties of metallic elements, their mixtures, i.e. alloys and their intermetallic compounds is known as metallurgy. It is further divided into two parts namely, non-ferrous metallurgy and ferrous metallurgy. The most important and widely used processes in this field are machining, casting, rolling, plating, heat treatment, thermal spraying, fabrication, forging, sintering, etc. This book presents the complex subject of metallurgy in the most comprehensible and easy to understand language. Different approaches, evaluations and methodologies have been included in it. This textbook will provide comprehensive knowledge to the readers.

Given below is the chapter wise description of the book:

Chapter 1- Metallurgy is an interdisciplinary subject of engineering and material science that studies metallic elements, compounds and alloys. It also deals with the production of metals and its methods. The subject is divided into ferrous metallurgy and non-ferrous metallurgy. This chapter will provide an integrated understanding of metallurgy.

Chapter 2- Two processes are applied to solve the problems of material balance arising during metal extraction. They are either on the basis of atomic/molecular weight of reactants or is based on the mass of compound. Stoichiometry combines these two notions and follows the principle of conservation of mass. The concepts of gas constant, thermochemistry, and material balance have been explored as well. The topics discussed in the chapter are of great importance to broaden the existing knowledge on metallurgy.

Chapter 3- Mineral processing separates the ore and the valuable minerals. The section elaborates on mineral processing methods such as roasting and calcination. In roasting, the concentrated ore is heated at a very high temperature. However, in calcination, the ore is subjected to high temperature in absence of air. All the diverse principles of mineral processing have been carefully analyzed in this chapter.

Chapter 4- Smelting, a part of extraction metallurgy, is used to produce base metals from ores. Smelting is combined with reduction for the extraction process. The reducing agent removes impurities and leaves behind the designated base metal. The section also lays impetus on iron-making and delves into the study of different smelting processes. The aspects elucidated in this chapter are of vital importance, and provide a better understanding of metallurgy.

Chapter 5- The chapter talks about furnaces such as blast furnace and cupola furnace and their related stoichiometries. A blast furnace is a high-temperature heating device that is used in the process of smelting. A cupola furnace is used to melt cast iron, and unlike a blast furnace, it could vary in size.

At the end, I would like to thank all those who dedicated their time and efforts for the successful completion of this book. I also wish to convey my gratitude towards my friends and family who supported me at every step.

Editor

1

Understanding Metallurgy

Metallurgy is an interdisciplinary subject of engineering and material science that studies metallic elements, compounds and alloys. It also deals with the production of metals and its methods. The subject is divided into ferrous metallurgy and non-ferrous metallurgy. This chapter will provide an integrated understanding of metallurgy.

Metallurgy

Metallurgy is a domain of materials science and engineering that studies the physical and chemical behaviour of metallicelements, their intermetallic compounds, and their mixtures, which are called alloys. Metallurgy is also the technology of metals: the way in which science is applied to the production of metals, and the engineering of metal components for usage in products for consumers and manufacturers. The production of metals involves the processing of ores to extract the metal they contain, and the mixture of metals, sometimes with other elements, to produce alloys. Metallurgy is distinguished from the craft of metalworking, although metalworking relies on metallurgy, as medicine relies on medical science, for technical advancement.

Smelting Gold in Nicaragua in the La Luz Gold Mine in Siuna and Bonanza about 1959. Smelting is a basic step in obtaining usable quantities of most metals.

Pouring Smelted Gold into an ingot at the La Luz Gold Mine in Siuna, Nicaragua about 1959.

Metallurgy is subdivided into *ferrous metallurgy* (sometimes also known as black metallurgy) and *non-ferrous metallurgy* or *colored metallurgy*. Ferrous metallurgy involves processes and alloys based on iron while non-ferrous metallurgy involves processes and alloys based on other metals. The production of ferrous metals accounts for 95 percent of world metal production.

Etymology and Pronunciation

The roots of *metallurgy* derive from Ancient Greek: *metallourgós*, "worker in metal", from *métallon*, "metal", "work".

The word was originally an alchemist's term for the extraction of metals from minerals, the ending *-urgy* signifying a process, especially manufacturing: it was discussed in this sense in the 1797 Encyclopaedia Britannica. In the late 19th century it was extended to the more general scientific study of metals, alloys, and related processes.

In English, the pronunciation is the more common one in the UK and Commonwealth. The pronunciation is the more common one in the USA, and is the first-listed variant in various American dictionaries (e.g., *Merriam-Webster Collegiate, American Heritage*).

History

Gold headband from Thebes 750–700 BC

The earliest recorded metal employed by humans appears to be gold, which can be found free or "native". Small amounts of natural gold have been found in Spanish caves used during the late Paleolithic period, c. 40,000 BC.Silver, copper, tin and meteoric iron can also be found in native form, allowing a limited amount of metalworking in early cultures. Egyptian weapons made from meteoric iron in about 3000 BC were highly prized as "daggers from heaven".

Certain metals, notably tin, lead and (at a higher temperature) copper, can be recovered from their ores by simply heating the rocks in a fire or blast furnace, a process known as smelting. The first evidence of this extractive metallurgy dates from the 5th and 6th millennia BC and was found in the archaeological sites of Majdanpek, Yarmovac and Plocnik, all three in Serbia. To date, the earliest evidence of copper smelting is found at the Belovode site,including a copper axe from 5500 BC belonging to the Vinča culture. Other signs of early metals are found from the third millennium BC in places like Palmela (Portugal), Los Millares (Spain), and Stonehenge (United Kingdom). However, the ultimate beginnings cannot be clearly ascertained and new discoveries are both continuous and ongoing.

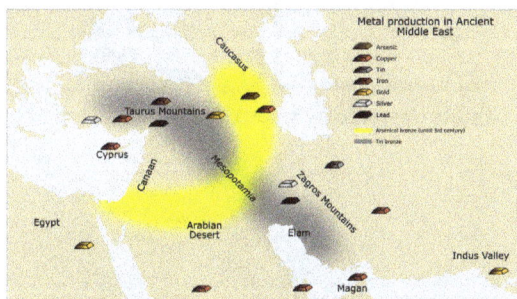

Mining areas of the ancient Middle East. Boxes colors: arsenic is in brown, copper in red, tin in grey, iron in reddish brown, gold in yellow, silver in white and lead in black. Yellow area stands for arsenic bronze, while grey area stands for tin bronze.

These first metals were single ones or as found. About 3500 BC, it was discovered that by combining copper and tin, a superior metal could be made, an alloy called bronze, representing a major technological shift known as the Bronze Age.

The extraction of iron from its ore into a workable metal is much more difficult than for copper or tin. The process appears to have been invented by the Hittites in about 1200 BC, beginning the Iron Age. The secret of extracting and working iron was a key factor in the success of the Philistines.

Historical developments in ferrous metallurgy can be found in a wide variety of past cultures and civilizations. This includes the ancient and medieval kingdoms and empires of the Middle East and Near East, ancient Iran, ancient Egypt, ancient Nubia, and Anatolia (Turkey), Ancient Nok, Carthage, the Greeks and Romans of ancient Europe, medieval Europe, ancient and medieval China, ancient and medieval India, ancient and medieval Japan, amongst others. Many applications, practices, and devices associated or involved in metallurgy were established in ancient China, such as the innovation of the blast furnace, cast iron, hydraulic-powered trip hammers, and double acting piston bellows.

A 16th century book by Georg Agricola called *De re metallica* describes the highly developed and complex processes of mining metal ores, metal extraction and metallurgy of the time. Agricola has been described as the "father of metallurgy".

Extraction

Furnace bellows operated by waterwheels, Yuan Dynasty, China.

Aluminium plant in Žiar nad Hronom (Central Slovakia)

Extractive metallurgy is the practice of removing valuable metals from an ore and refining the extracted raw metals into a purer form. In order to convert a metal oxide or sulphide to a purer metal, the ore must be reduced physically, chemically, or electrolytically.

Extractive metallurgists are interested in three primary streams: feed, concentrate (valuable metal oxide/sulphide), and tailings (waste). After mining, large pieces of the ore feed are broken through crushing and/or grinding in order to obtain particles small enough where each particle is either mostly valuable or mostly waste. Concentrating the particles of value in a form supporting separation enables the desired metal to be removed from waste products.

Mining may not be necessary if the ore body and physical environment are conducive to leaching. Leaching dissolves minerals in an ore body and results in an enriched solution. The solution is collected and processed to extract valuable metals.

Ore bodies often contain more than one valuable metal. Tailings of a previous process may be used as a feed in another process to extract a secondary product from the original ore. Additionally, a concentrate may contain more than one valuable metal. That concentrate would then be processed to separate the valuable metals into individual constituents.

Alloys

Casting bronze

Common engineering metals include aluminium, chromium, copper, iron, magnesium, nickel, titanium and zinc. These are most often used as alloys. Much effort has been placed on understanding the iron-carbon alloy system, which includes steels and cast irons. Plain carbon steels (those that contain essentially only carbon as an alloying element) are used in low-cost, high-strength applications where weight and corrosion are not a problem. Cast irons, including ductile iron, are also part of the iron-carbon system.

Stainless steel or galvanized steel are used where resistance to corrosion is important. Aluminium alloys and magnesium alloys are used for applications where strength and lightness are required.

Copper-nickel alloys (such as Monel) are used in highly corrosive environments and for non-magnetic applications. Nickel-based superalloys like Inconel are used in high-temperature applications such as gas turbines, turbochargers, pressure vessels, and heat exchangers. For extremely high temperatures, single crystal alloys are used to minimize creep.

Production

In production engineering, metallurgy is concerned with the production of metallic components for use in consumer or engineering products. This involves the production of alloys, the shaping, the heat treatment and the surface treatment of the product. The task of the metallurgist is to achieve balance between material properties such as cost, weight, strength, toughness, hardness, corrosion, fatigue resistance, and performance in temperature extremes. To achieve this goal, the operating environment must be carefully considered. In a saltwater environment, ferrous metals and some aluminium alloys corrode quickly. Metals exposed to cold or cryogenic conditions may endure a ductile to brittle transition and lose their toughness, becoming more brittle and prone to cracking. Metals under continual cyclic loading can suffer from metal fatigue. Metals under constant stress at elevated temperatures can creep.

Metalworking Processes

Metals are shaped by processes such as:

- casting – molten metal is poured into a shaped mold.

- forging – a red-hot billet is hammered into shape.

- rolling – a billet is passed through successively narrower rollers to create a sheet.

- laser cladding – metallic powder is blown through a movable laser beam (e.g. mounted on a NC 5-axis machine). The resulting melted metal reaches a substrate to form a melt pool. By moving the laser head, it is possible to stack the tracks and build up a three-dimensional piece.

- extrusion – a hot and malleable metal is forced under pressure through a die, which shapes it before it cools.

- sintering – a powdered metal is heated in a non-oxidizing environment after being compressed into a die.

- machining – lathes, milling machines, and drills cut the cold metal to shape.

- fabrication – sheets of metal are cut with guillotines or gas cutters and bent and welded into structural shape.

- 3D printing – Sintering or melting powder metal in a very small point on a moving 'print head' moving in 3D space to make any object to shape.

Cold-working processes, in which the product's shape is altered by rolling, fabrication or other processes while the product is cold, can increase the strength of the product by a process called work hardening. Work hardening creates microscopic defects in the metal, which resist further changes of shape.

Various forms of casting exist in industry and academia. These include sand casting, investment casting (also called the "lost wax process"), die casting, and continuous casting.

Heat Treatment

Metals can be heat-treated to alter the properties of strength, ductility, toughness, hardness and/or resistance to corrosion. Common heat treatment processes include annealing, precipitation strengthening, quenching, and tempering. The annealing process softens the metal by heating it and then allowing it to cool very slowly, which gets rid of stresses in the metal and makes the grain structure large and soft-edged so that when the metal is hit or stressed it dents or perhaps bends, rather than breaking; it is also easier to sand, grind, or cut annealed metal. Quenching is the process of cooling a high-carbon steel very quickly after heating, thus "freezing" the steel's molecules in the very hard martensite form, which makes the metal harder. There is a balance between hardness and toughness in any steel; the harder the steel, the less tough or impact-resistant it is, and the more impact-resistant it is, the less hard it is. Tempering relieves stresses in the metal that were caused by the hardening process; tempering makes the metal less hard while making it better able to sustain impacts without breaking.

Often, mechanical and thermal treatments are combined in what are known as thermo-mechanical treatments for better properties and more efficient processing of materials. These processes are common to high-alloy special steels, superalloys and titanium alloys.

Plating

Electroplating is a chemical surface-treatment technique. It involves bonding a thin layer of another metal such as gold, silver, chromium or zinc to the surface of the product. It is used to reduce corrosion as well as to improve the product's aesthetic appearance.

Thermal Spraying

Thermal spraying techniques are another popular finishing option, and often have better high temperature properties than electroplated coatings.

Microstructure

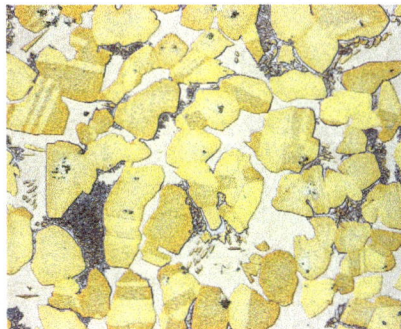

Metallography allows the metallurgist to study the microstructure of metals.

Metallurgists study the microscopic and macroscopic properties using metallography, a technique invented by Henry Clifton Sorby. In metallography, an alloy of interest is ground flat and polished to a mirror finish. The sample can then be etched to reveal the microstructure and macrostructure of the metal. The sample is then examined in an optical or electron microscope, and the image contrast provides details on the composition, mechanical properties, and processing history.

Crystallography, often using diffraction of x-rays or electrons, is another valuable tool available to the modern metallurgist. Crystallography allows identification of unknown materials and reveals the crystal structure of the sample. Quantitative crystallography can be used to calculate the amount of phases present as well as the degree of strain to which a sample has been subjected.

Conferences

EMC, the European Metallurgical Conference has developed to the most important networking business event dedicated to the non-ferrous metals industry in Europe. From the start of the conference sequence in 2001 at Friedrichshafen it was host of the most relevant metallurgists from all countries of the world. The European Metallurgical Conference is organized by GDMB Society of Metallurgists and Miners.

Characterization of Natural Reserves of Metals

The natural reserve of a metal is called "ore". Ore is an aggregate of minerals. A mineral in an inorganic compound in which elements are mixed in stoichiometric proportion, for example Al_2O_3 is a mineral in which 2 moles of aluminum are combined with 1.5 moles of oxygen gas. An ore of any metal contains valuable mineral and gangue minerals. Valuable mineral is the mineral which is used to produce metal.

In the ore, metal grade is important.

$$\text{Metal grade of an ore} = \frac{\text{Amount of metal in ore}}{\text{Amount of ore.}} \times 100$$

It must be noted very clearly that ore does not contain metal but metal in the ore is in the form of a mineral. Metal grade is used to characterize an ore reserve. For example metal grade of iron in pure Fe_2O_3 is 70%. If iron ore contains 80% Fe_2O_3, then iron grade of ore in 56%. This means that 44% of the ore is waste both in terms of solid and oxygen of the valuable mineral. In the following table metal grade of certain ores, valuable minerals etc. are given

We Note the Following Form the Above Table:

Metal	Ore	Valuable mineral	Metal grade (%)
Al	Bauxite	Al_2O_3	17.4%.
Ti	Ilmenite	$Ti\,O_2$	36%

Cu	Sulphide	$CuFeS_2$	2 to 3%.
Fe	Hematite	Fe_2O_3	56% 64%
Ni	Sulphide ore	Ni_3S_2	2.3%
Pb	Sulphide ore	PbS	5.5%

i) Metal in the valuable mineral is chemically combined with either oxygen or sulphur.

ii) Metal grade in sulphide ore is very low as compared with oxide ores.

iii) Low grade of any ore means large production of wastage. Thus waste production is a part of metal production from natural reserves.

Metal Extraction Requirement

The chemical combination of metal with sulphur or oxygen in a mineral is accompanied with high heat of formation

Mineral	Heat of formation ΔH_f^c (k cal /kg mole)
Fe_2O_3	198×10^3
Al_2O_3	380×10^3
Cu_2S	19×10^3
ZnS	44×10^3
PbS	22×10^3
Fe_3O_4	26×10^3
$Ti O_2$	218×10^3
MgO	142×10^3

Thus large amount of energy would be required for example to produce Fe or Al from their respective minerals. Two basic methods of metal extraction are:

Pyrometallurgical Extraction

This method depends on whether oxide ore or sulphide ore is employed for metal extraction. In the extraction first step is to upgrade the metal grade by employing mineral beneficiation technologies. The concentrate is then mixed with a suitable reducing agent and the mixture is followed by smelting to separate gangue minerals form metal. Here large amount of energy is required to facilitate separation of metal from gangue.

Energy source and its consumption are important. For the sulphide ore the following route is employed.

Sulphide is first converted to oxide and then following by smelting. Zinc and lead are produced by this route.

In another route sulphides are used to obtain by matte by matte smelting. It is used to produce copper.

Hydrometallurgical Extraction

Hydrometallurgical extraction is mainly suited to lean ores.In the hydrometallurgical extraction metal is brought in to solution by leaching, which is then followed either by hydrogen reduction or electrolytic reduction to obtain metal.

This route is used to produce Al and Mg.

Energy Requirement for Metal Production

Pyrometallurgical extraction of metals requires large amount of energy as shown in the following table:

Metal	Gross energy Requirement (mj/kg metal)
Titanium	360
Aluminium	220
Nickel	110
Zinc	50
Copper	40-50
Steel	15-20

Metal grade of the ore is important. Low grade ore will consume more energy. Additional energy is required for mining and mineral processing to remove gangue mineral.

This is particularly true for copper and nickel extraction: for example 0.5% Cu grade of ore will consume 250 Mj/kg of energy which will decreases to 50 MJ/kg when copper grade of ore is 2%. Similarly a nickel grade of 0.5% will require 375 MJ/ kg energy which will decrease to 150 MJ / kg when nickel grade is 2%. Metal grade of ore is very important.

Sources of Energy

In pyrometallurgical extraction thermal energy is required. Fossil fuel is the source of energy. Fossil fuels are the non renewable source of energy and hence their optimum utilization is important. Further thermal energy from fossil fuel is derived by combustion which leads to production of products of combustion like CO, CO_2, NO_x etc. Large energy requirement demands higher consumption of fossil fuel.

Thus optimum utilization of fossil fuel in pyrometallurgical extraction is important from the point of view of conservation of natural resources and cleanliness of the environment.

Environmental Issues

The production of metals from natural reserves results in the formation of emissions, unwanted solids, liquids and gases like CO, CO_2, NO_x SO_2, SO_3 etc directly (during mining and processing) and indirectly(associated with the consumption of raw materials and utilities), for example in the generation of electric power. In the supply chains of metal needs, mineral resource extraction and processing are particularly critical stages for the potential release of gas, liquid and solid emissions.Sulphurus gases can cause acidic rains. NO_x is a group of different gases formed due to different levels of nitrogen and oxygen. Commonly formed gases are nitrogen dioxide and nitric oxide.NO_x is given off in many forms such as smog or particles.NO_x cobributes to global warming, hampers the growth of plants and forms acid rains. It is harmful to humans as well. It can cause nausea, irritated eyes and major respiratory problems.

Environment impact of the process depends on metal grade of ore, electrical energy source, fuel types and material transport as well as processing technology. As higher grade ore reserves of metal decrease, there will be a dramatic effect on the energy consumption and accompanying greenhouse emissions from metal production processes.

The environment impacts for cradle-to-gate metal production (cradle-to-gate is an assessment of a partial production life cycle from resource extraction to the factory gate) are given in the following table:

Environmental Impacts for "Cradle-to-gate" Metal Production

Metal	Process	GER * (MJ/ kg)	GWP$(kg CO_2 e/ kg)	AP#(kg SO_2e/kg)	SWB## (kg/kg)
Nickel	flash furnace smelting and sherritt Gordon refining	114	11.4	0.130	65
Copper	pressure acid leaching Sx/Ew Smelting /converting and electro –refining Heap leaching and Sx/ Ew	194	16.2	-	351
		33	3.3	0.040	64
		64	6.2	-	125
Lead	Lead blast furnace Imperial smelting process	20	2.1	0.022	14.8
		32	3.2	0.035	15.9
Zinc	Electrolytic process Imperial smelting process	48	4.6	0.055	29.3

Aluminum	Bayer refining ,Hall- Heroult smelting	36	3.3	0.036	15.4
		211	22.4	0.131	4.5
Titanium	Beecher and Kroll process	361	35.7	0.230	16.9
Steel	Integrated route (BF and BOF)	23	2.3	0.020	2.4
Stainless steel	Electric furnace and Argon–Oxygen decarburization	75	6.8	0.051	6.4

* GER: Gross energy requirement

$ GWP: Global warming potential.

AP: Acidification potential.

SWB: solid waste burden.

Energy balance is a very powerful tool to address the issues of saving natural resources.

What can be done more beyond Fuel Savings?

Facts about use of natural energy resources

Fossil fuel based energy availability is associated with

a) Discharge of CO_2 in the environment and other harmful gases like SO_2, NO_x etc. 1 kg mole of carbon discharges 1 kg mole of CO_2 in the environment. In other words 1 kg carbon produces 3.7 kg CO_2.

b) Large amount of heat is carried away by the products of combustion. In metal extraction processes at high temperatures, products like liquid metal, slag, gases and coke carry a large fraction of heat.

Sensible heat is the issue of concern. In this connection it is important to understand quality of heat. Quality of heat is decided by the temperature. Higher is the temperature of discharged product, higher is the quality of heat.

Units of Measurement

The adoption of standards has varied greatly as regards unit in different parts of the world. The following table lists some of the commonly used set of fundamental units from which all other units can be derived.

Quantity	Absolute units FPS CGS MKS SI	Engg.Units English
Mass	Pound gram kg kg	Slug
Length	ft cm m m	ft
Time	Sec Sec Sec s	sec
Amount of substance	Lb.mole g.mole kgmole mole	Lb. mole
Temperature	°F °C °K(ork) K	°F

FPS stands for feet, pound and second. It is british system

CGS stands for centimeter, gram and second

MKS stands for meter, kilogram and second

SI stands for international system of units. Extension of MKS system

Derived Units and Quantities

Physical quantities which can be derived from other physical quantities are called derived quantities.

Derived units: The units of physical quantities which can be expressed in terms of fundamental units, for example area, volume etc.

Lets us Derive

(i) unit of force

Force =mass x acceleration.

In SI unit force $= \dfrac{kgm}{s^2} = 1N$ (Newton).

In CGS unit force $= \dfrac{gcm}{s^2} = 1\,dyne$

In MKS unit force is expensed in Newton (N)

In FPS systems force is $\propto \dfrac{bft}{s^2} = 1\,poundal$

(ii) Energy $=mu^2$

we can substitute the units of m and u to derive unit of energy.

In SI system $E = kg\dfrac{m2}{s2} = 1Joule$.

In CGS system $E = g\dfrac{Cm2}{s2} = 1erg$.

1 joule = 10^7 erg.

In FPS system $E = Lb\ ft^2/s^2 = ft \cdot poundal$.

1British thermal unit (Btu) $= 778$ ft. Lbf $= 1054.2$ Joule $= 252$ cal

1kcal $= 1000$cal.$= 3.968$ Btu

(iii) power (watts or W).

$$1w = 1\frac{kgm^2}{s^3} = j/s$$

1horse power $= 550$ ft.lbf /s $= 746W$

1KW $= 1000W = 3414$ Btu/hr.

1KW- hr $= 860$ kcal $= 3414$ Btu.

Composition of a Mixture

It is expressed in terms of mole fraction or mass fraction. If mixture conatins n, moles of component 1, n_2 moles of component 2, n_3 moles of component of 3.; then mole fraction of ith component is

$$x_i = \frac{ni}{\Sigma_i n_i}$$

Similarly mass fraction of ith component in the mixture is given by.

$$Y_i = \frac{m_i}{\Sigma_i m_i}$$

Derivation of unit of universal constant (R)

$$R = \frac{p_o v_o}{T_0} = \frac{latms \times 22415cm^3}{gmole \times 273ok} = 82.10\frac{cm^3.atm}{gmole\ ok}$$

In CGS Unit

$$PO = 1.013125 \times 10^6 dynes\ cm^{-2}, To = 273k, Vo = 22451\frac{cm^3}{g.mole}$$

$$R = \frac{1.01325 \times 10^6 \times 22415}{273} = 8.319 \times 10^7\frac{ergs}{g\ mole\ k}$$

$Sin\ ce\ 1\ J = 107\ erg$ and $1\ cal = 4.184\ J$

$\therefore R = 1.988cal / (g\ mole\ k)$.

In SI units

$$Po = 1.013125 * 10^5 \text{pascal}, To = 273K \text{ and } Vo = 0.022451\frac{m^3}{g.mol}$$

$$R = 8.314\frac{J}{g.mole\ k}$$

Similarly in M K S system $R = 8.314\frac{J}{g.mole\ k}$.

Current =Ampere (A). In solids current consists of electron flow. In electrolyte solutios. most of the current flow by motion of ionic species for example cu^+ or $Na+$

1coulomb is unit of charge: flow of 1A/s.

SI unit of electrical potential is volt. Volt is the potential in which the charge of 1 coulomb experiences a force of 1 Newtron.

SI unit of resistance is ohm. Ohm is defined as the resistance which permits flow of 1 A current under an imposed electrical potential difference of 1V.

Some Basic Equations of Electrical Flow are:

$$V = R,I$$

$$P = I.V = I^2R \quad P = power$$

$$t = time$$

$W = tP = I^2\ Rt$ .W=energy measured in joules and power in watts

A Faraday is one mole of electrons.

1 faraday = 96500 coulomb. One faraday will discharge one gram equivalent of ions.

The liberation of one g equivalent of any metal consumes 96500 coulombs of electricity

How many gram moles of Al^{3+} ions could be discharged in one minute by 1.9×10^4 A current, if no loss of current occur.

In one minute a current of 1.9×10^4 A will carry 1.9×10^4 coulombs of electricity.

Gram moles of Al deposited $= \dfrac{1.14\times10^6}{3\times96500} = 3.94$

Concentration of Solids in Slurry

Many metallurgical processes have feed and/or product streams that consist of mixtures of solids and liquids. These mixtures are called slurries.

The relationship between wt % solid (%x) and specific gravity of solid phase (ρ_s) and that of slurry (ρ_m) when water is used as a medium can be obtained:

Volume of slurry = Volume of solid + Volume of water. Consider 1 kg slurry with %x as solids weight percent, then

$$\frac{1}{\rho m} = \frac{\%x}{100_{\rho s}} + \frac{(100-\%x)}{100\rho w}$$

$$\rho_w = \text{density of water}$$

$$\rho = \text{density of solid}$$

$$\rho_m = \text{density of mixture (solid + water)}$$

$$(\text{Wt percent solid})\%x = \frac{100_{\rho s}(\rho_m - 1000)}{\rho_m(\rho_s - 1000)} \quad (1)$$

$$\text{Volume }\%\text{slurry} = \%x \frac{\rho m}{\rho s}$$

Mass Flow Rate of Dry Solid in Slurry (M)

$$= \frac{\text{volumetric flow rate} \times \text{slurry deunty} \times \%x}{100}$$

$$M = \frac{F\rho m \%x}{100} \text{ kg/hr} \quad (2)$$

F is volume flow rate in m3/hr.

By1 and 2

$$M = \frac{F\rho s(\rho m - 1000)}{(\rho s - 1000)} \quad (3)$$

2

Concepts Related to Metallurgy

Two processes are applied to solve the problems of material balance arising during metal extraction. They are either on the basis of atomic/molecular weight of reactants or is based on the mass of compound. Stoichiometry combines these two notions and follows the principle of conservation of mass. The concepts of gas constant, thermo-chemistry, and material balance have been explored as well. The topics discussed in the chapter are of great importance to broaden the existing knowledge on metallurgy.

Stoichiometry

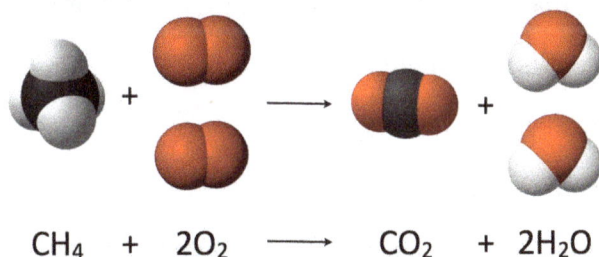

$$CH_4 \ + \ 2O_2 \ \longrightarrow \ CO_2 \ + \ 2H_2O$$

Stoichiometry is the calculation of relative quantities of reactants and products in chemical reactions.

Stoichiometry is founded on the law of conservation of mass where the total mass of the reactants equals the total mass of the products leading to the insight that the relations among quantities of reactants and products typically form a ratio of positive integers. This means that if the amounts of the separate reactants are known, then the amount of the product can be calculated. Conversely, if one reactant has a known quantity and the quantity of product can be empirically determined, then the amount of the other reactants can also be calculated.

This is illustrated in the image here, where the balanced equation is:

$$CH_4 + 2\,O_2 \ \ CO_2 + 2\,H_2O.$$

Here, one molecule of methane reacts with two molecules of oxygen gas to yield one molecule of carbon dioxide and two molecules of water. Stoichiometry measures these quantitative relationships, and is used to determine the amount of products/reactants that are produced/needed in a given reaction. Describing the quantitative relationships among substances as they participate in chemical reactions is known as *reaction stoichiometry*. In the example above, reaction stoichiometry measures the relationship between the methane and oxygen as they react to form carbon dioxide and water.

Because of the well known relationship of moles to atomic weights, the ratios that are arrived at by stoichiometry can be used to determine quantities by weight in a reaction described by a balanced equation. This is called *composition stoichiometry*.

Gas stoichiometry deals with reactions involving gases, where the gases are at a known temperature, pressure, and volume and can be assumed to be ideal gases. For gases, the volume ratio is ideally the same by the ideal gas law, but the mass ratio of a single reaction has to be calculated from the molecular masses of the reactants and products. In practice, due to the existence of isotopes, molar masses are used instead when calculating the mass ratio.

Definition

A stoichiometric amount or stoichiometric ratio of a reagent is the optimum amount or ratio where, assuming that the reaction proceeds to completion:

1.   All of the reagent is consumed

2.   There is no deficiency of the reagent

3.   There is no excess of the reagent.

Stoichiometry rests upon the very basic laws that help to understand it better, i.e., law of conservation of mass, the law of definite proportions (i.e., the law of constant composition), the law of multiple proportions and the law of reciprocal proportions. In general, chemical reactions combine in definite ratios of chemicals. Since chemical reactions can neither create nor destroy matter, nor transmute one element into another, the amount of each element must be the same throughout the overall reaction. For example, the number of atoms of a given element X on the reactant side must equal the number of atoms of that element on the product side, whether or not all of those atoms are actually involved in a reaction.

Chemical reactions, as macroscopic unit operations, consist of simply a very large number of elementary reactions, where a single molecule reacts with another molecule. As the reacting molecules (or moieties) consist of a definite set of atoms in an integer ratio, the ratio between reactants in a complete reaction is also in integer ratio. A reaction may consume more than one molecule, and the stoichiometric number counts this number, defined as positive for products (added) and negative for reactants (removed).

Different elements have a different atomic mass, and as collections of single atoms, molecules have a definite molar mass, measured with the unit mole (6.02×10^{23} individual molecules, Avogadro's constant). By definition, carbon-12 has a molar mass of 12 g/mol. Thus, to calculate the stoichiometry by mass, the number of molecules required for each reactant is expressed in moles and multiplied by the molar mass of each to give the mass of each reactant per mole of reaction. The mass ratios can be calculated by dividing each by the total in the whole reaction.

Elements in their natural state are mixtures of isotopes of differing mass, thus atomic masses and thus molar masses are not exactly integers. For instance, instead of an exact 14:3 proportion, 17.04 kg of ammonia consists of 14.01 kg of nitrogen and 3 × 1.01 kg of hydrogen, because natural nitrogen includes a small amount of nitrogen-15, and natural hydrogen includes hydrogen-2 (deuterium).

A stoichiometric reactant is a reactant that is consumed in a reaction, as opposed to a catalytic reactant, which is not consumed in the overall reaction because it reacts in one step and is regenerated in another step.

Converting Grams to Moles

Stoichiometry is not only used to balance chemical equations but also used in conversions, i.e., converting from grams to moles using molar mass as the conversion factor, or from grams to milliliters using density. For example, to find the amount of NaCl (sodium chloride) in 2.00 g, one would do the following:

$$\frac{2.00 \text{ g NaCl}}{58.44 \text{ g NaCl mol}^{-1}} = 0.034 \text{ mol}$$

In the above example, when written out in fraction form, the units of grams form a multiplicative identity, which is equivalent to one (g/g = 1), with the resulting amount in moles (the unit that was needed), as shown in the following equation,

$$\left(\frac{2.00 \text{ g NaCl}}{1}\right)\left(\frac{1 \text{ mol NaCl}}{58.44 \text{ g NaCl}}\right) = 0.034 \text{ mol}$$

Molar Proportion

Stoichiometry is often used to balance chemical equations (reaction stoichiometry). For example, the two diatomic gases, hydrogen and oxygen, can combine to form a liquid, water, in an exothermic reaction, as described by the following equation:

$$2 \text{ H}_2 + \text{O}_2 \rightarrow 2 \text{ H}_2\text{O}$$

Reaction stoichiometry describes the 2:1:2 ratio of hydrogen, oxygen, and water molecules in the above equation.

The molar ratio allows for conversion between moles of one substance and moles of another. For example, in the following reaction:

$$2 \; CH_3OH + 3 \; O_2 \rightarrow 2 \; CO_2 + 4 \; H_2O$$

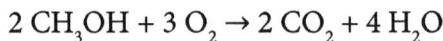

the amount of water that will be produced by the combustion of 0.27 moles of CH3OH is obtained using the molar ratio between CH_3OH and H_2O of 2 to 4.

$$\left(\frac{0.27 \; \text{mol} \; CH_3OH}{1} \right) \left(\frac{4 \; \text{mol} \; H_2O}{2 \; \text{mol} \; CH_3OH} \right) = 0.54 \; \text{mol} \; H_2O$$

The term stoichiometry is also often used for the molar proportions of elements in stoichiometric compounds (composition stoichiometry). For example, the stoichiometry of hydrogen and oxygen in H_2O is 2:1. In stoichiometric compounds, the molar proportions are whole numbers.

Determining Amount of Product

Stoichiometry can also be used to find the quantity of a product yielded by a reaction. If a piece of solid copper (Cu) were added to an aqueous solution of silver nitrate $(AgNO_3)$, the silver (Ag) would be replaced in a single displacement reaction forming aqueous copper(II) nitrate $(Cu(NO_3)_2)$ and solid silver. How much silver is produced if 16.00 grams of Cu is added to the solution of excess silver nitrate?

The following steps would be used:

1. Write and balance the equation

2. Mass to moles: Convert grams of Cu to moles of Cu

3. Mole ratio: Convert moles of Cu to moles of Ag produced

4. Mole to mass: Convert moles of Ag to grams of Ag produced

The complete balanced equation would be:

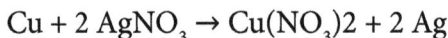

$$Cu + 2 \; AgNO_3 \rightarrow Cu(NO_3)2 + 2 \; Ag$$

For the mass to mole step, the mass of copper (16.00 g) would be converted to moles of copper by dividing the mass of copper by its molecular mass: 63.55 g/mol.

$$\left(\frac{16.00 \; \text{g} \; Cu}{1} \right) \left(\frac{1 \; \text{mol} \; Cu}{63.55 \; \text{g} \; Cu} \right) = 0.2518 \; \text{mol} \; Cu$$

Now that the amount of Cu in moles (0.2518) is found, we can set up the mole ratio. This is found by looking at the coefficients in the balanced equation: Cu and Ag are in a 1:2 ratio.

$$\left(\frac{0.2518 \; \text{mol} \; Cu}{1} \right) \left(\frac{2 \; \text{mol} \; Ag}{1 \; \text{mol} \; Cu} \right) = 0.5036 \; \text{mol} \; Ag$$

Now that the moles of Ag produced is known to be 0.5036 mol, we convert this amount to grams of Ag produced to come to the final answer:

$$\left(\frac{0.5036 \text{ mol Ag}}{1}\right)\left(\frac{107.87 \text{ g Ag}}{1 \text{ mol Ag}}\right) = 54.32 \text{ g Ag}$$

This set of calculations can be further condensed into a single step:

$$m_{Ag} = \left(\frac{16.00 \text{ g Cu}}{1}\right)\left(\frac{1 \text{ mol Cu}}{63.55 \text{ g Cu}}\right)\left(\frac{2 \text{ mol Ag}}{1 \text{ mol Cu}}\right)\left(\frac{107.87 \text{ g Ag}}{1 \text{ mol Ag}}\right) = 54.32 \text{ g}$$

Further Examples

For propane (C_3H_8) reacting with oxygen gas (O_2), the balanced chemical equation is:

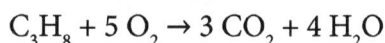

$$C_3H_8 + 5 \text{ } O_2 \rightarrow 3 \text{ } CO_2 + 4 \text{ } H_2O$$

The mass of water formed if 120 g of propane (C_3H_8) is burned in excess oxygen is then

$$m_{H_2O} = \left(\frac{120. \text{ g } C_3H_8}{1}\right)\left(\frac{1 \text{ mol } C_3H_8}{44.09 \text{ g } C_3H_8}\right)\left(\frac{4 \text{ mol } H_2O}{1 \text{ mol } C_3H_8}\right)\left(\frac{18.02 \text{ g } H_2O}{1 \text{ mol } H_2O}\right) = 196 \text{ g}$$

Stoichiometric Ratio

Stoichiometry is also used to find the right amount of one reactant to "completely" react with the other reactant in a chemical reaction – that is, the stoichiometric amounts that would result in no leftover reactants when the reaction takes place. An example is shown below using the thermite reaction,

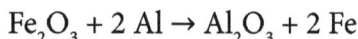

$$Fe_2O_3 + 2 \text{ Al} \rightarrow Al_2O_3 + 2 \text{ Fe}$$

This equation shows that 1 mole of iron(III) oxide and 2 moles of aluminum will produce 1 mole of aluminium oxide and 2 moles of iron. So, to completely react with 85.0 g of iron(III) oxide (0.532 mol), 28.7 g (1.06 mol) of aluminium are needed.

$$m_{Al} = \left(\frac{85.0 \text{ g } Fe_2O_3}{1}\right)\left(\frac{1 \text{ mol } Fe_2O_3}{159.7 \text{ g } Fe_2O_3}\right)\left(\frac{2 \text{ mol Al}}{1 \text{ mol } Fe_2O_3}\right)\left(\frac{26.98 \text{ g Al}}{1 \text{ mol Al}}\right) = 28.7 \text{ g}$$

Limiting Reagent and Percent Yield

The limiting reagent is the reagent that limits the amount of product that can be formed and is completely consumed when the reaction is complete. An excess reactant is a reactant that is left over once the reaction has stopped due to the limiting reactant being exhausted.

Consider the equation of roasting lead(II) sulfide (PbS) in oxygen (O_2) to produce lead(II) oxide (PbO) and sulfur dioxide (SO_2):

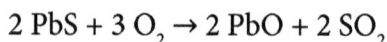

$$2 \text{ PbS} + 3 \text{ O}_2 \rightarrow 2 \text{ PbO} + 2 \text{ SO}_2$$

To determine the theoretical yield of lead(II) oxide if 200.0 g of lead(II) sulfide and 200.0 g of oxygen are heated in an open container:

$$m_{\text{PbO}} = \left(\frac{200.0 \text{ g PbS}}{1}\right)\left(\frac{1 \text{ mol PbS}}{239.27 \text{ g PbS}}\right)\left(\frac{2 \text{ mol PbO}}{2 \text{ mol PbS}}\right)\left(\frac{223.2 \text{ g PbO}}{1 \text{ mol PbO}}\right) = 186.6 \text{ g}$$

$$m_{\text{PbO}} = \left(\frac{200.0 \text{ g O}_2}{1}\right)\left(\frac{1 \text{ mol O}_2}{32.00 \text{ g O}_2}\right)\left(\frac{2 \text{ mol PbO}}{3 \text{ mol O}_2}\right)\left(\frac{223.2 \text{ g PbO}}{1 \text{ mol PbO}}\right) = 930.0 \text{ g}$$

Because a lesser amount of PbO is produced for the 200.0 g of PbS, it is clear that PbS is the limiting reagent.

In reality, the actual yield is not the same as the stoichiometrically-calculated theoretical yield. Percent yield, then, is expressed in the following equation:

$$\text{percent yield} = \frac{\text{actual yield}}{\text{theoretical yield}}$$

If 170.0 g of lead(II) oxide is obtained, then the percent yield would be calculated as follows:

$$\text{percent yield} = \frac{170.0 \text{ g PbO}}{186.6 \text{ g PbO}} = 91.12\%$$

Example

Consider the following reaction, in which iron(III) chloride reacts with hydrogen sulfide to produce iron(III) sulfide and hydrogen chloride:

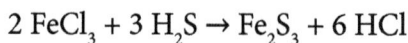

$$2 \text{ FeCl}_3 + 3 \text{ H}_2\text{S} \rightarrow \text{Fe}_2\text{S}_3 + 6 \text{ HCl}$$

Suppose 90.0 g of $FeCl_3$ reacts with 52.0 g of H_2S. To find the limiting reagent and the mass of HCl produced by the reaction, we could set up the following equations:

$$m_{\text{HCl}} = \left(\frac{90.0 \text{ g FeCl}_3}{1}\right)\left(\frac{1 \text{ mol FeCl}_3}{162 \text{ g FeCl}_3}\right)\left(\frac{6 \text{ mol HCl}}{2 \text{ mol FeCl}_3}\right)\left(\frac{36.5 \text{ g HCl}}{1 \text{ mol HCl}}\right) = 60.8 \text{ g}$$

$$m_{\text{HCl}} = \left(\frac{52.0 \text{ g H}_2\text{S}}{1}\right)\left(\frac{1 \text{ mol H}_2\text{S}}{34.1 \text{ g H}_2\text{S}}\right)\left(\frac{6 \text{ mol HCl}}{3 \text{ mol H}_2\text{S}}\right)\left(\frac{36.5 \text{ g HCl}}{1 \text{ mol HCl}}\right) = 111 \text{ g}$$

Thus, the limiting reagent is $FeCl_3$ and the amount of HCl produced is 60.8 g.

To find what mass of excess reagent (H_2S) remains after the reaction, we would set up the calculation to find out how much H_2S reacts completely with the 90.0 g $FeCl_3$:

$$m_{H_2S} = \left(\frac{90.0 \text{ g FeCl}_3}{1}\right)\left(\frac{1 \text{ mol FeCl}_3}{162 \text{ g FeCl}_3}\right)\left(\frac{3 \text{ mol H}_2S}{2 \text{ mol FeCl}_3}\right)\left(\frac{34.1 \text{ g H}_2S}{1 \text{ mol H}_2S}\right) = 28.4 \text{ g reacted}$$

By subtracting this amount from the original amount of H_2S, we can come to the answer:

$$52.0 \text{ g H}_2S - 28.4 \text{ g H}_2S = 23.6 \text{ g H}_2S \text{ excess}$$

Different Stoichiometries in Competing Reactions

Often, more than one reaction is possible given the same starting materials. The reactions may differ in their stoichiometry. For example, the methylation of benzene (C_6H_6), through a Friedel–Crafts reaction using $AlCl_3$ as a catalyst, may produce singly methylated ($C_6H_5CH_3$), doubly methylated ($C_6H_4(CH_3)_2$), or still more highly methylated ($C_6H_{6-n}(CH_3)_n$) products, as shown in the following example,

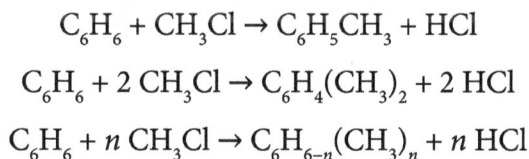

$$C_6H_6 + CH_3Cl \rightarrow C_6H_5CH_3 + HCl$$

$$C_6H_6 + 2 CH_3Cl \rightarrow C_6H_4(CH_3)_2 + 2 HCl$$

$$C_6H_6 + n CH_3Cl \rightarrow C_6H_{6-n}(CH_3)_n + n HCl$$

In this example, which reaction takes place is controlled in part by the relative concentrations of the reactants.

Stoichiometric Coefficient

In lay terms, the *stoichiometric coefficient* (or *stoichiometric number* in the IUPAC nomenclature) of any given component is the number of molecules that participate in the reaction as written.

For example, in the reaction $CH_4 + 2 O_2 \rightarrow CO_2 + 2 H_2O$, the stoichiometric coefficient of CH_4 is −1, the stoichiometric coefficient of O_2 is −2, for CO_2 it would be +1 and for H_2O it is +2.

In more technically precise terms, the stoichiometric coefficient in a chemical reaction-system of the ith component is defined as

$$v_i = \frac{\Delta N_i}{\Delta \xi}$$

or

$$\Delta N_i = v_i \Delta \xi$$

where N_i is the number of molecules of i, and ξ is the progress variable or extent of reaction.

The extent of reaction ξ can be regarded as [the amount of] a real (or hypothetical) product, one molecule of which produced each time the reaction event occurs. It is the extensive quantity describing the progress of a chemical reaction equal to the number of chemical transformations, as indicated by the reaction equation on a molecular scale, divided by the Avogadro constant (in essence, it is the amount of chemical transformations). The change in the extent of reaction is given by $d\xi = dn_B/v_B$, where v_B is the stoichiometric number of any reaction entity B (reactant or product) and n_B is the corresponding amount.

The stoichiometric coefficient v_i represents the degree to which a chemical species participates in a reaction. The convention is to assign negative coefficients to *reactants* (which are consumed) and positive ones to *products*. However, any reaction may be viewed as going in the reverse direction, and all the coefficients then change sign (as does the free energy). Whether a reaction actually *will* go in the arbitrarily selected forward direction or not depends on the amounts of the substances present at any given time, which determines the kinetics and thermodynamics, i.e., whether equilibrium lies to the *right* or the *left*.

In reaction mechanisms, stoichiometric coefficients for each step are always integers, since elementary reactions always involve whole molecules. If one uses a composite representation of an overall reaction, some may be rationalfractions. There are often chemical species present that do not participate in a reaction; their stoichiometric coefficients are therefore zero. Any chemical species that is regenerated, such as a catalyst, also has a stoichiometric coefficient of zero.

The simplest possible case is an isomerization

$$A \rightarrow B$$

in which $v_B = 1$ since one molecule of B is produced each time the reaction occurs, while $v_A = -1$ since one molecule of A is necessarily consumed. In any chemical reaction, not only is the total mass conserved but also the numbers of atoms of each kind are conserved, and this imposes corresponding constraints on possible values for the stoichiometric coefficients.

There are usually multiple reactions proceeding simultaneously in any natural reaction system, including those in biology. Since any chemical component can participate in several reactions simultaneously, the stoichiometric coefficient of the ith component in the kth reaction is defined as

$$v_{ik} = \frac{\partial N_i}{\partial \xi_k}$$

so that the total (differential) change in the amount of the ith component is

$$dN_i = \sum_k v_{ik} d\xi_k.$$

Extents of reaction provide the clearest and most explicit way of representing compositional change, although they are not yet widely used.

With complex reaction systems, it is often useful to consider both the representation of a reaction system in terms of the amounts of the chemicals present $\{N_i\}$ (state variables), and the representation in terms of the actual compositional degrees of freedom, as expressed by the extents of reaction $\{\xi_k\}$. The transformation from a vector expressing the extents to a vector expressing the amounts uses a rectangular matrix whose elements are the stoichiometric coefficients $[v_{ik}]$.

The maximum and minimum for any ξ_k occur whenever the first of the reactants is depleted for the forward reaction; or the first of the "products" is depleted if the reaction as viewed as being pushed in the reverse direction. This is a purely kinematic restriction on the reaction simplex, a hyperplane in composition space, or Nspace, whose dimensionality equals the number of *linearly-independent* chemical reactions. This is necessarily less than the number of chemical components, since each reaction manifests a relation between at least two chemicals. The accessible region of the hyperplane depends on the amounts of each chemical species actually present, a contingent fact. Different such amounts can even generate different hyperplanes, all sharing the same algebraic stoichiometry.

In accord with the principles of chemical kinetics and thermodynamic equilibrium, every chemical reaction is *reversible*, at least to some degree, so that each equilibrium point must be an interior point of the simplex. As a consequence, extrema for the ξs will not occur unless an experimental system is prepared with zero initial amounts of some products.

The number of *physically*-independent reactions can be even greater than the number of chemical components, and depends on the various reaction mechanisms. For example, there may be two (or more) reaction *paths* for the isomerism above. The reaction may occur by itself, but faster and with different intermediates, in the presence of a catalyst.

The (dimensionless) "units" may be taken to be molecules or moles. Moles are most commonly used, but it is more suggestive to picture incremental chemical reactions in terms of molecules. The Ns and ξs are reduced to molar units by dividing by Avogadro's number. While dimensional mass units may be used, the comments about integers are then no longer applicable.

Stoichiometry Matrix

In complex reactions, stoichiometries are often represented in a more compact form called the stoichiometry matrix. The stoichiometry matrix is denoted by the symbol N.

If a reaction network has n reactions and m participating molecular species then the stoichiometry matrix will have correspondingly m rows and n columns.

For example, consider the system of reactions shown below:

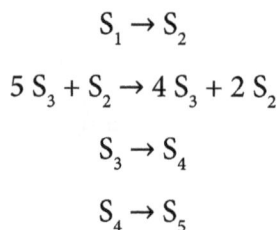

$$S_1 \rightarrow S_2$$

$$5\,S_3 + S_2 \rightarrow 4\,S_3 + 2\,S_2$$

$$S_3 \rightarrow S_4$$

$$S_4 \rightarrow S_5$$

This systems comprises four reactions and five different molecular species. The stoichiometry matrix for this system can be written as:

$$
N = \begin{bmatrix}
-1 & 0 & 0 & 0 \\
1 & 1 & 0 & 0 \\
0 & -1 & -1 & 0 \\
0 & 0 & 1 & -1 \\
0 & 0 & 0 & 1
\end{bmatrix}
$$

where the rows correspond to S_1, S_2, S_3, S_4 and S_5, respectively. Note that the process of converting a reaction scheme into a stoichiometry matrix can be a lossy transformation, for example, the stoichiometries in the second reaction simplify when included in the matrix. This means that it is not always possible to recover the original reaction scheme from a stoichiometry matrix.

Often the stoichiometry matrix is combined with the rate vector, v, and the species vector, S to form a compact equation describing the rates of change of the molecular species:

$$\frac{dS}{dt} = N \cdot v.$$

Gas Stoichiometry

Gas stoichiometry is the quantitative relationship (ratio) between reactants and products in a chemical reaction with reactions that produce gases. Gas stoichiometry applies when the gases produced are assumed to be ideal, and the temperature, pressure, and volume of the gases are all known. The ideal gas law is used for these calculations. Often, but not always, the standard temperature and pressure (STP) are taken as 0 °C and 1 bar and used as the conditions for gas stoichiometric calculations.

Gas stoichiometry calculations solve for the unknown volume or mass of a gaseous product or reactant. For example, if we wanted to calculate the volume of gaseous NO_2 produced from the combustion of 100 g of NH_3, by the reaction:

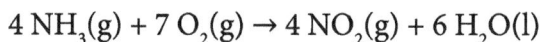

$$4\,NH_3(g) + 7\,O_2(g) \rightarrow 4\,NO_2(g) + 6\,H_2O(l)$$

we would carry out the following calculations:

$$100g\,NH_3 \cdot \frac{1mol\,NH_3}{17.034g\,NH_3} = 5.871mol\,NH_3$$

There is a 1:1 molar ratio of NH_3 to NO_2 in the above balanced combustion reaction, so 5.871 mol of NO_2 will be formed. We will employ the ideal gas law to solve for the volume at 0 °C (273.15 K) and 1 atmosphere using the gas law constant of $R = 0.08206$ L·atm·K^{-1}·mol^{-1} :

$$PV = nRT$$
$$V = \frac{nRT}{P}$$
$$= \frac{5.871 \cdot 0.08206 \cdot 273.15}{1}$$
$$= 131.597 L\,NO_2$$

Gas stoichiometry often involves having to know the molar mass of a gas, given the density of that gas. The ideal gas law can be re-arranged to obtain a relation between the density and the molar mass of an ideal gas:

$$\rho = \frac{m}{V} \quad \text{and} \quad n = \frac{m}{M}$$

and thus:

$$\rho = \frac{MP}{RT}$$

where:

- P = absolute gas pressure

- V = gas volume

- n = amount (measured in moles)

- R = universal ideal gas law constant

- T = absolute gas temperature

- ρ = gas density at T and P

- m = mass of gas

- M = molar mass of gas

Stoichiometric air-to-fuel Ratios of Common Fuels

In the combustion reaction, oxygen reacts with the fuel, and the point where exactly all oxygen is consumed and all fuel burned is defined as the stoichiometric point. With more oxygen (overstoichiometric combustion), some of it stays unreacted. Likewise, if the combustion is incomplete due to lack of sufficient oxygen, fuel remains unreacted. (Unreacted fuel may also remain because of slow combustion or insufficient mixing of fuel and oxygen – this is not due to stoichiometry). Different hydrocarbon fuels have different contents of carbon, hydrogen and other elements, thus their stoichiometry varies.

Fuel	Ratio by mass	Ratio by volume	Percent fuel by mass
Gasoline	14.7 : 1	—	6.8%
Natural gas	17.2 : 1	9.7 : 1	5.8%
Propane (LP)	15.67 : 1	23.9 : 1	6.45%
Ethanol	9 : 1	—	11.1%
Methanol	6.47 : 1	—	15.6%
n-Butanol	11.2 : 1	—	8.2%
Hydrogen	34.3 : 1	2.39 : 1	2.9%
Diesel	14.5 : 1	—	6.8%
Methane	17.19 : 1	9.52 : 1	5.5%

Gasoline engines can run at stoichiometric air-to-fuel ratio, because gasoline is quite volatile and is mixed (sprayed or carburetted) with the air prior to ignition. Diesel engines, in contrast, run lean, with more air available than simple stoichiometry would require. Diesel fuel is less volatile and is effectively burned as it is injected, leaving less time for evaporation and mixing. Thus, it would form soot (black smoke) at stoichiometric ratio.

Gas Constant

Values of R	Units ($VPT^{-1}n^{-1}$)
8.314 4598(48)	kg m^2 s^{-2} K^{-1} mol^{-1}
8.3144598(48)	J K^{-1} mol^{-1}
	kJ K^{-1} kmol^{-1}
8.3144598(48)×10^7	erg K^{-1} mol^{-1}

8.3144598(48)×10⁻³	amu (km/s)² K⁻¹
8.3144598(48)	m³ Pa K⁻¹ mol⁻¹
8.3144598(48)×10⁺⁶	cm³ Pa K⁻¹ mol⁻¹
8.3144598(48)	L kPa K⁻¹ mol⁻¹
8.3144598(48)×10³	cm³ kPa K⁻¹ mol⁻¹
8.3144598(48)×10⁻⁶	m³ MPa K⁻¹ mol⁻¹
8.3144598(48)	cm³ MPa K⁻¹ mol⁻¹
8.3144598(48)×10⁻⁵	m³ bar K⁻¹ mol⁻¹
8.3144598(48)×10⁻²	L bar K⁻¹ mol⁻¹
83.144598(48)	cm³ bar K⁻¹ mol⁻¹
62.363577(36)	L Torr K⁻¹ mol⁻¹
1.9872036(11)×10⁻³	kcal K⁻¹ mol⁻¹
8.2057338(47)×10⁻⁵	m³ atm K⁻¹ mol⁻¹
8.2057338(47)×10⁻²	L atm K⁻¹ mol⁻¹
82.057338(47)	cm³ atm K⁻¹ mol⁻¹

The gas constant (also known as the molar, universal, or ideal gas constant, denoted by the symbol R or $\mathcal{R}$) is a physical constant which is featured in many fundamental equations in the physical sciences, such as the ideal gas law and the Nernst equation.

It is equivalent to the Boltzmann constant, but expressed in units of energy (i.e. the pressure-volume product) per temperature increment per *mole* (rather than energy per temperature increment per *particle*). The constant is also a combination of the constants from Boyle's law, Charles's law, Avogadro's law, and Gay-Lussac's law.

Physically, the gas constant is the constant of proportionality that happens to relate the energy scale in physics to the temperature scale, when a mole of particles at the stated temperature is being considered. Thus, the value of the gas constant ultimately derives from historical decisions and accidents in the setting of the energy and temperature scales, plus similar historical setting of the value of the molar scale used for the counting of particles. The last factor is not a consideration in the value of the Boltzmann constant, which does a similar job of equating linear energy and temperature scales.

The gas constant value is

8.3144598(48) J·mol⁻¹·K⁻¹

The two digits in parentheses are the uncertainty (standard deviation) in the last two digits of the value. The relative uncertainty is 5.7×10^{-7}. Some have suggested that it might be appropriate to name the symbol R the Regnault constant in honour of the Frenchchemist Henri Victor Regnault, whose accurate experimental data were used to calculate the early value of the constant; however, the exact reason for the original representation of the constant by the letter R is elusive.

The gas constant occurs in the ideal gas law, as follows:

$$PV = nRT = mR_{\text{specific}} T$$

where P is the absolute pressure (SI unit pascals), V is the volume of gas (SI unit cubic metres), n is the amount of gas (SI unit moles), m is the mass (SI unit kilograms) contained in V, and T is the thermodynamic temperature (SI unit kelvin). The gas constant is expressed in the same physical units as molar entropy and molar heat capacity.

Dimensions of R

From the general equation $PV = nRT$ we get:

$$R = \frac{PV}{nT}$$

where P is pressure, V is volume, n is number of moles of a given substance, and T is temperature.

As pressure is defined as force per unit area, the gas equation can also be written as:

$$R = \frac{\dfrac{\text{force}}{\text{area}} \times \text{volume}}{\text{amount} \times \text{temperature}}$$

Area and volume are (length)² and (length)³ respectively. Therefore:

$$R = \frac{\dfrac{\text{force}}{(\text{length})^2} \times (\text{length})^3}{\text{amount} \times \text{temperature}} = \frac{\text{force} \times \text{length}}{\text{amount} \times \text{temperature}}$$

Since force × length = work:

$$R = \frac{\text{work}}{\text{amount} \times \text{temperature}}$$

The physical significance of R is work per degree per mole. It may be expressed in any set of units representing work or energy (such as joules), other units representing degrees of temperature (such as degrees Celsius or Fahrenheit), and any system of units designating a mole or a similar pure number that allows an equation of macroscopic mass and fundamental particle numbers in a system, such as an ideal gas.

Instead of a mole the constant can be expressed by considering the normal cubic meter.

Otherwise, we can also say that:

$$\text{force} = \frac{\text{mass} \times \text{length}}{(\text{time})^2}$$

Therefore, we can write "R" as:

$$R = \frac{\text{mass} \times \text{length}^2}{\text{amount} \times \text{temperature} \times (\text{time})^2}$$

And so, in SI base units:

$$R = 8.3144598(48) \text{ kg m}^2 \text{ mol}^{-1} \text{ K}^{-1} \text{ s}^{-2}.$$

Relationship with the Boltzmann Constant

The Boltzmann constant k_B (often abbreviated k) may be used in place of the gas constant by working in pure particle count, N, rather than amount of substance, n, since

$$R = N_A k_B,$$

where N_A is the Avogadro constant. For example, the ideal gas law in terms of Boltzmann's constant is

$$PV = k_B NT.$$

where N is the number of particles (molecules in this case), or to generalize to an inhomogeneous system the local form holds:

$$P = k_B nT.$$

where n is the number density.

Measurement

As of 2006, the most precise measurement of R is obtained by measuring the speed of sound $c_a(p, T)$ in argon at the temperature T of the triple point of water (used to define the kelvin) at different pressures p, and extrapolating to the zero-pressure limit $c_a(0, T)$. The value of R is then obtained from the relation

$$c_a(0, T) = \sqrt{\frac{\gamma_0 RT}{A_r(\text{Ar})M_u}},$$

where:

- γ_0 is the heat capacity ratio ($\frac{5}{3}$ for monatomic gases such as argon);

- T is the temperature, $T_{TPW} = 273.16$ K by definition of the kelvin;

- $A_r(\text{Ar})$ is the relative atomic mass of argon and $M_u = 10^{-3}$ kg mol^{-1}.

Specific Gas Constant

$R_{specific}$ for dry air	Units
287.058	$J\ kg^{-1}\ K^{-1}$
53.3533	$ft\ lbf\ lb^{-1}\ °R^{-1}$
1,716.49	$ft\ lbf\ slug^{-1}\ °R^{-1}$
Based on a mean molar mass for dry air of 28.9645 g/mol.	

The specific gas constant of a gas or a mixture of gases ($R_{specific}$) is given by the molar gas constant divided by the molar mass (M) of the gas or mixture.

$$R_{specific} = \frac{R}{M}$$

Just as the ideal gas constant can be related to the Boltzmann constant, so can the specific gas constant by dividing the Boltzmann constant by the molecular mass of the gas.

$$R_{specific} = \frac{k_B}{m}$$

Another important relationship comes from thermodynamics. Mayer's relation relates the specific gas constant to the specific heats for a calorically perfect gas and a thermally perfect gas.

$$R_{specific} = c_p - c_v$$

where c_p is the specific heat for a constant pressure and c_v is the specific heat for a constant volume.

It is common, especially in engineering applications, to represent the specific gas constant by the symbol R. In such cases, the universal gas constant is usually given a different symbol such as $\bar{R}$ to distinguish it. In any case, the context and/or units of the gas constant should make it clear as to whether the universal or specific gas constant is being referred to.

U.S. Standard Atmosphere

The U.S. Standard Atmosphere, 1976 (USSA1976) defines the gas constant R^* as:

$$R^* = 8.31432 \times 10^3\ N\ m\ kmol^{-1}\ K^{-1}.$$

Note the use of kilomole units resulting in the factor of 1,000 in the constant. The USSA1976 acknowledges that this value is not consistent with the cited values for the Avogadro constant and the Boltzmann constant. This disparity is not a significant departure from accuracy, and USSA1976 uses this value of $R*$ for all the calculations of the standard atmosphere. When using the ISO value of R, the calculated pressure increases by only 0.62 pascal at 11 kilometers (the equivalent of a difference of only 17.4 centimeters or 6.8 inches) and an increase of 0.292 Pa at 20 km (the equivalent of a difference of only 0.338 m or 13.2 in).

Excess and Limiting Reactants

The exact amount of the reactants can be determined by the stoichiometry of the reaction. In many cases one or more reactants is supplied in excess to complete the reaction. Reactant present in the smallest stoichiometric amount is called the limiting reactant. Reactants supplied in excess of the limiting reactant are called excess reactants.

Oxidation- Reduction Reactions.

Oxidation: Loss of electrons from an atom

Reduction: Gain of electrons by an atom

Oxidation and reduction always occurs simultaneously so that electrons are conserved.

Cementation reaction

Thermochemistry

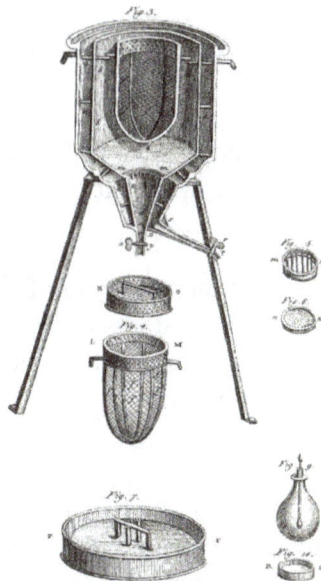

The world's first ice-calorimeter, used in the winter of 1782-83, by Antoine Lavoisier and Pierre-Simon Laplace, to determine the heat evolved in various chemical changes; calculations which were based on Joseph Black's prior discovery of latent heat. These experiments mark the foundation of thermochemistry.

Thermochemistry is the study of the energy and heat associated with chemical reactions and/or physical transformations. A reaction may release or absorb energy, and a phase change may do the same, such as in melting and boiling. Thermochemistry focuses on these energy changes, particularly on the system's energy exchange with its surroundings. Thermochemistry is useful in predicting reactant and product quantities throughout the course of a given reaction. In combination with entropy determinations, it is also used to predict whether a reaction is spontaneous or non-spontaneous, favorable or unfavorable.

Endothermic reactions absorb heat, while exothermic reactions release heat. Thermochemistry coalesces the concepts of thermodynamics with the concept of energy in the form of chemical bonds. The subject commonly includes calculations of such quantities as heat capacity, heat of combustion, heat of formation, enthalpy, entropy, free energy, and calories.

History

Thermochemistry rests on two generalizations. Stated in modern terms, they are as follows:

1. Lavoisier and Laplace's law (1780): The energy change accompanying any transformation is equal and opposite to energy change accompanying the reverse process.

2. Hess' law (1840): The energy change accompanying any transformation is the same whether the process occurs in one step or many.

These statements preceded the first law of thermodynamics (1845) and helped in its formulation.

Lavoisier, Laplace and Hess also investigated specific heat and latent heat, although it was Joseph Black who made the most important contributions to the development of latent energy changes.

Gustav Kirchhoff showed in 1858 that the variation of the heat of reaction is given by the difference in heat capacity between products and reactants: $d\Delta H / dT = \Delta C_p$. Integration of this equation permits the evaluation of the heat of reaction at one temperature from measurements at another temperature.

Calorimetry

The measurement of heat changes is performed using calorimetry, usually an enclosed chamber within which the change to be examined occurs. The temperature of the chamber is monitored either using a thermometer or thermocouple, and the temperature plotted against time to give a graph from which fundamental quantities can be calculated.

Modern calorimeters are frequently supplied with automatic devices to provide a quick read-out of information, one example being the differential scanning calorimeter (DSC).

Systems

Several thermodynamic definitions are very useful in thermochemistry. A system is the specific portion of the universe that is being studied. Everything outside the system is considered the surroundings or environment. A system may be:

- a (completely) isolated system which can exchange neither energy nor matter with the surroundings, such as an insulated bomb calorimeter

- a thermally isolated system which can exchange mechanical work but not heat or matter, such as an insulated closed piston or balloon

- a mechanically isolated system which can exchange heat but not mechanical work or matter, such as an uninsulated bomb calorimeter

- a closed system which can exchange energy but not matter, such as an uninsulated closed piston or balloon

- an open system which it can exchange both matter and energy with the surroundings, such as a pot of boiling water

Processes

A system undergoes a process when one or more of its properties changes. A process relates to the change of state. An isothermal (same-temperature) process occurs when temperature of the system remains constant. An isobaric (same-pressure) process occurs when the pressure of the system remains constant. A process is adiabatic when no heat exchange occurs.

First Law of Thermodynamics

The first law of thermodynamics is a version of the law of conservation of energy, adapted for thermodynamic systems. The law of conservation of energy states that the total energy of an isolated system is constant; energy can be transformed from one form to another, but can be neither created nor destroyed. The first law is often formulated by stating that the change in the internal energy of a closed system is equal to the amount of heat supplied to the system, minus the amount of work done by the system on its surroundings. Equivalently, perpetual motion machines of the first kind are impossible.

History

Investigations into the nature of heat and work and their relationship began with the invention of the first engines used to extract water from mines. Improvements to such engines so as to increase their efficiency and power output came first from mechanics that worked with such machines but only slowly advanced the art. Deeper investigations that placed those on a mathematical and physics basis came later.

The first law of thermodynamics was developed empirically over about half a century. The first full statements of the law came in 1850 from Rudolf Clausius and from William Rankine; Rankine's statement is less distinct relative to Clausius'. A main aspect of the struggle was to deal with the previously proposed caloric theory of heat.

In 1840, Germain Hess stated a conservation law for the so-called 'heat of reaction' for chemical reactions. His law was later recognized as a consequence of the first law of thermodynamics, but Hess's statement was not explicitly concerned with the relation between energy exchanges by heat and work.

According to Truesdell (1980), Julius Robert von Mayer in 1841 made a statement that meant that "in a process at constant pressure, the heat used to produce expansion is universally interconvertible with work", but this is not a general statement of the first law.

Original Statements: the Thermodynamic Approach

The original nineteenth century statements of the first law of thermodynamics appeared in a conceptual framework in which transfer of energy as heat was taken as a primitive notion, not defined or constructed by the theoretical development of the framework, but rather presupposed as prior to it and already accepted. The primitive notion of heat was taken as empirically established, especially through calorimetry regarded as a subject in its own right, prior to thermodynamics. Jointly primitive with this notion of heat were the notions of empirical temperature and thermal equilibrium. This framework also took as primitive the notion of transfer of energy as work. This framework did not presume a concept of energy in general, but regarded it as derived or synthesized from the prior notions of heat and work. By one author, this framework has been called the "thermodynamic" approach.

The first explicit statement of the first law of thermodynamics, by Rudolf Clausius in 1850, referred to cyclic thermodynamic processes.

> *In all cases in which work is produced by the agency of heat, a quantity of heat is consumed which is proportional to the work done; and conversely, by the expenditure of an equal quantity of work an equal quantity of heat is produced.*

Clausius also stated the law in another form, referring to the existence of a function of state of the system, the internal energy, and expressed it in terms of a differential equation for the increments of a thermodynamic process. This equation may be described as follows:

In a thermodynamic process involving a closed system, the increment in the internal energy is equal to the difference between the heat accumulated by the system and the work done by it.

Because of its definition in terms of increments, the value of the internal energy of a system is not uniquely defined. It is defined only up to an arbitrary additive constant of integration, which can be adjusted to give arbitrary reference zero levels. This non-uniqueness is in keeping with the abstract mathematical nature of the internal energy. The internal energy is customarily stated relative to a conventionally chosen standard reference state of the system.

The concept of internal energy is considered by Bailyn to be of "enormous interest". Its quantity cannot be immediately measured, but can only be inferred, by differencing actual immediate measurements. Bailyn likens it to the energy states of an atom, that were revealed by Bohr's energy relation $h\nu = E_{n''} - E_{n'}$. In each case, an unmeasurable quantity (the internal energy, the atomic energy level) is revealed by considering the difference of measured quantities (increments of internal energy, quantities of emitted or absorbed radiative energy).

Conceptual Revision: the Mechanical Approach

In 1907, George H. Bryan wrote about systems between which there is no transfer of matter (closed systems): "Definition. When energy flows from one system or part of a system to another otherwise than by the performance of mechanical work, the energy so transferred is called *heat*." This definition may be regarded as expressing a conceptual revision, as follows. This was systematically expounded in 1909 by Constantin Carathéodory, whose attention had been drawn to it by Max Born. Largely through Born's influence, this revised conceptual approach to the definition of heat came to be preferred by many twentieth-century writers. It might be called the "mechanical approach".

Energy can also be transferred from one thermodynamic system to another in association with transfer of matter. Born points out that in general such energy transfer is not resolvable uniquely into work and heat moieties. In general, when there is transfer of energy associated with matter transfer, work and heat transfers can be distinguished only when they pass through walls physically separate from those for matter transfer.

The "mechanical" approach postulates the law of conservation of energy. It also postulates that energy can be transferred from one thermodynamic system to another adiabatically as work, and that energy can be held as the internal energy of a thermodynamic system. It also postulates that energy can be transferred from one thermodynamic system to another by a path that is non-adiabatic, and is unaccompanied by matter transfer. Initially, it "cleverly" (according to Bailyn) refrains from labelling as 'heat' such non-adiabatic, unaccompanied transfer of energy. It rests on the primitive notion of *walls*, especially adiabatic walls and non-adiabatic walls, defined as follows. Temporarily, only for purpose of this definition, one can prohibit transfer of energy as work across a wall of interest. Then walls of interest fall into two classes, (a) those such that arbitrary systems separated by them remain independently in their own previously established respective states of internal thermodynamic equilibrium; they are defined as adiabatic; and (b) those without such independence; they are defined as non-adiabatic.

This approach derives the notions of transfer of energy as heat, and of temperature, as theoretical developments, not taking them as primitives. It regards calorimetry as a derived theory. It has an early origin in the nineteenth century, for example in the work of Helmholtz, but also in the work of many others.

Conceptually Revised Statement: According to the Mechanical Approach

The revised statement of the first law postulates that a change in the internal energy of a system due to any arbitrary process, that takes the system from a given initial thermodynamic state to a given final equilibrium thermodynamic state, can be determined through the physical existence, for those given states, of a reference process that occurs purely through stages of adiabatic work.

The revised statement is then

> *For a closed system, in any arbitrary process of interest that takes it from an initial to a final state of internal thermodynamic equilibrium, the change of internal energy is the same as that for a reference adiabatic work process that links those two states. This is so regardless of the path of the process of interest, and regardless of whether it is an adiabatic or a non-adiabatic process. The reference adiabatic work process may be chosen arbitrarily from amongst the class of all such processes.*

This statement is much less close to the empirical basis than are the original statements, but is often regarded as conceptually parsimonious in that it rests only on the concepts of adiabatic work and of non-adiabatic processes, not on the concepts of transfer of energy as heat and of empirical temperature that are presupposed by the original statements. Largely through the influence of Max Born, it is often regarded as theoretically preferable because of this conceptual parsimony. Born particularly observes that the revised approach avoids thinking in terms of what he calls the "imported engineering" concept of heat engines.

Basing his thinking on the mechanical approach, Born in 1921, and again in 1949, proposed to revise the definition of heat. In particular, he referred to the work of Constantin Carathéodory, who had in 1909 stated the first law without defining quantity of heat. Born's definition was specifically for transfers of energy without transfer of matter, and it has been widely followed in textbooks (examples:). Born observes that a transfer of matter between two systems is accompanied by a transfer of internal energy that cannot be resolved into heat and work components. There can be pathways to other systems, spatially separate from that of the matter transfer, that allow heat and work transfer independent of and simultaneous with the matter transfer. Energy is conserved in such transfers.

Description

The first law of thermodynamics for a closed system was expressed in two ways by Clausius. One way referred to cyclic processes and the inputs and outputs of the system, but did not refer to increments in the internal state of the system. The other way referred to an incremental change in the internal state of the system, and did not expect the process to be cyclic.

A cyclic process is one that can be repeated indefinitely often, returning the system to its initial state. Of particular interest for single cycle of a cyclic process are the net work done, and the net heat taken in (or 'consumed', in Clausius' statement), by the system.

In a cyclic process in which the system does net work on its surroundings, it is observed to be physically necessary not only that heat be taken into the system, but also, importantly, that some heat leave the system. The difference is the heat converted by the cycle into work. In each repetition of a cyclic process, the net work done by the system, measured in mechanical units, is proportional to the heat consumed, measured in calorimetric units.

The constant of proportionality is universal and independent of the system and in 1845 and 1847 was measured by James Joule, who described it as the *mechanical equivalent of heat.*

In a non-cyclic process, the change in the internal energy of a system is equal to net energy added as heat to the system minus the net work done by the system, both being measured in mechanical units. Taking ΔU as a change in internal energy, one writes

$$\Delta U = Q - W$$

where Q denotes the net quantity of heat supplied to the system by its surroundings and W denotes the net work done by the system. This sign convention is implicit in Clausius' statement of the law given above. It originated with the study of heat engines that produce useful work by consumption of heat.

Often nowadays, however, writers use the IUPAC convention by which the first law is formulated with work done on the system by its surroundings having a positive sign. With this now often used sign convention for work, the first law for a closed system may be written:

$$\Delta U = Q + W \text{ (sign convention of IUPAC).}$$

This convention follows physicists such as Max Planck, and considers all net energy transfers to the system as positive and all net energy transfers from the system as negative, irrespective of any use for the system as an engine or other device.

When a system expands in a fictive quasistatic process, the work done by the system on the environment is the product, $P \, dV$, of pressure, P, and volume change, dV, whereas the work done *on* the system is $-P \, dV$. Using either sign convention for work, the change in internal energy of the system is:

$$dU = \delta Q - P dV \text{ (quasi-static process)},$$

where δQ denotes the infinitesimal increment of heat supplied to the system from its surroundings.

Work and heat are expressions of actual physical processes of supply or removal of energy, while the internal energy U is a mathematical abstraction that keeps account of the exchanges of energy that befall the system. Thus the term heat for Q means "that amount of energy added or removed by conduction of heat or by thermal radiation", rather than referring to a form of energy within the system. Likewise, the term work energy for W means "that amount of energy gained or lost as the result of work". Internal energy is a property of the system whereas work done and heat supplied are not. A significant result of this distinction is that a given internal energy change ΔU can be achieved by, in principle, many combinations of heat and work.

Various Statements of the Law for Closed Systems

The law is of great importance and generality and is consequently thought of from several points of view. Most careful textbook statements of the law express it for closed systems. It is stated in several ways, sometimes even by the same author.

For the thermodynamics of closed systems, the distinction between transfers of energy as work and as heat is central. For the thermodynamics of open systems, such a distinction is beyond the scope, but some limited comments are made on it in the section below headed 'First law of thermodynamics for open systems'.

There are two main ways of stating a law of thermodynamics, physically or mathematically. They should be logically coherent and consistent with one another.

An example of a physical statement is that of Planck:

> It is in no way possible, either by mechanical, thermal, chemical, or other devices, to obtain perpetual motion, i.e. it is impossible to construct an engine which will work in a cycle and produce continuous work, or kinetic energy, from nothing.

This physical statement is restricted neither to closed systems nor to systems with states that are strictly defined only for thermodynamic equilibrium; it has meaning also for open systems and for systems with states that are not in thermodynamic equilibrium.

An example of a mathematical statement is that of Crawford (1963):

> For a given system we let ΔE^{kin} = large-scale mechanical energy, ΔE^{pot} = large-scale potential energy, and ΔE^{tot} = total energy. The first two quantities are specifiable in terms of appropriate mechanical variables, and by definition
>
> $$E^{tot} = E^{kin} + E^{pot} + U.$$
>
> For any finite process, whether reversible or irreversible,
>
> $$\Delta E^{tot} = \Delta E^{kin} + \Delta E^{pot} + \Delta U.$$
>
> The first law in a form that involves the principle of conservation of energy more generally is
>
> $$\Delta E^{tot} = Q + W.$$
>
> Here Q and W are heat and work added, with no restrictions as to whether the process is reversible, quasistatic, or irreversible.

This statement by Crawford, for W, uses the sign convention of IUPAC, not that of Clausius. Though it does not explicitly say so, this statement refers to closed systems, and to internal energy U defined for bodies in states of thermodynamic equilibrium, which possess well-defined temperatures.

The history of statements of the law for closed systems has two main periods, before and after the work of Bryan (1907), of Carathéodory (1909), and the approval of Carathéodory's work given by Born (1921). The earlier traditional versions of the law for closed systems are nowadays often considered to be out of date.

Carathéodory's celebrated presentation of equilibrium thermodynamics refers to closed systems, which are allowed to contain several phases connected by internal walls of various kinds of impermeability and permeability (explicitly including walls that are permeable only to heat). Carathéodory's 1909 version of the first law of thermodynamics was stated in an axiom which refrained from defining or mentioning temperature or quantity of heat transferred. That axiom stated that the internal energy of a phase in equilibrium is a function of state, that the sum of the internal energies of the phases is the total internal energy of the system, and that the value of the total internal energy of the system is changed by the amount of work done adiabatically on it, considering work as a form of energy. The article considered this statement to be an expression of the law of conservation of energy for such systems. This version is nowadays widely accepted as authoritative, but is stated in slightly varied ways by different authors.

Such statements of the first law for closed systems assert the existence of internal energy as a function of state defined in terms of adiabatic work. Thus heat is not defined calorimetrically or as due to temperature difference. It is defined as a residual difference between change of internal energy and work done on the system, when that work does not account for the whole of the change of internal energy and the system is not adiabatically isolated.

The 1909 Carathéodory statement of the law in axiomatic form does not mention heat or temperature, but the equilibrium states to which it refers are explicitly defined by variable sets that necessarily include "non-deformation variables", such as pressures, which, within reasonable restrictions, can be rightly interpreted as empirical temperatures, and the walls connecting the phases of the system are explicitly defined as possibly impermeable to heat or permeable only to heat.

According to Münster (1970), "A somewhat unsatisfactory aspect of Carathéodory's theory is that a consequence of the Second Law must be considered at this point [in the statement of the first law], i.e. that it is not always possible to reach any state 2 from any other state 1 by means of an adiabatic process." Münster instances that no adiabatic process can reduce the internal energy of a system at constant volume. Carathéodory's paper asserts that its statement of the first law corresponds exactly to Joule's experimental arrangement, regarded as an instance of adiabatic work. It does not point out that Joule's experimental arrangement performed essentially irreversible work, through friction of paddles in a liquid, or passage of electric current through a resistance inside the system, driven by motion of a coil and inductive heating, or by an external current source, which can access the system only by the passage of electrons, and so is not strictly adiabatic, because electrons are a form of matter, which cannot penetrate adiabatic walls. The paper goes on to base its main argument on the possibility of quasi-static adiabatic work, which is essentially reversible. The paper asserts that it will avoid reference to Carnot cycles, and then proceeds to base its argument on cycles of forward and backward quasi-static adiabatic stages, with isothermal stages of zero magnitude.

Sometimes the concept of internal energy is not made explicit in the statement.

Sometimes the existence of the internal energy is made explicit but work is not explicitly mentioned in the statement of the first postulate of thermodynamics. Heat supplied is then defined as the residual change in internal energy after work has been taken into account, in a non-adiabatic process.

A respected modern author states the first law of thermodynamics as "Heat is a form of energy", which explicitly mentions neither internal energy nor adiabatic work. Heat is defined as energy transferred by thermal contact with a reservoir, which has a temperature, and is generally so large that addition and removal of heat do not alter its temperature. A current student text on chemistry defines heat thus: "*heat* is the exchange of thermal energy between a system and its surroundings caused by a temperature difference." The author then explains how heat is defined or measured by calorimetry, in terms of heat capacity, specific heat capacity, molar heat capacity, and temperature.

A respected text disregards the Carathéodory's exclusion of mention of heat from the statement of the first law for closed systems, and admits heat calorimetrically defined along with work and internal energy. Another respected text defines heat exchange as determined by temperature difference, but also mentions that the Born (1921) version is "completely rigorous". These versions follow the traditional approach that is now considered out of date, exemplified by that of Planck.

Evidence for the First Law of Thermodynamics for Closed Systems

The first law of thermodynamics for closed systems was originally induced from empirically observed evidence, including calorimetric evidence. It is nowadays, however, taken to provide the definition of heat via the law of conservation of energy and the definition of work in terms of changes in the external parameters of a system. The original discovery of the law was gradual over a period of perhaps half a century or more, and some early studies were in terms of cyclic processes.

The following is an account in terms of changes of state of a closed system through compound processes that are not necessarily cyclic. This account first considers processes for which the first law is easily verified because of their simplicity, namely adiabatic processes(in which there is no transfer as heat) and adynamic processes (in which there is no transfer as work).

Adiabatic Processes

In an adiabatic process, there is transfer of energy as work but not as heat. For all adiabatic process that takes a system from a given initial state to a given final state, irrespective of how the work is done, the respective eventual total quantities of energy transferred as work are one and the same, determined just by the given initial and final states. The work done on the system is defined and measured by changes in mechanical or quasi-mechanical variables external to the system. Physically, adiabatic transfer of energy as work requires the existence of adiabatic enclosures.

For instance, in Joule's experiment, the initial system is a tank of water with a paddle wheel inside. If we isolate the tank thermally, and move the paddle wheel with a pulley and a weight, we can relate the increase in temperature with the distance descended by the mass. Next, the system is returned to its initial state, isolated again, and the same amount of work is done on the tank using different devices (an electric motor, a chemical battery, a spring,...). In every case, the amount of work can be measured independently. The return to the initial state is not conducted by doing adiabatic work on the system. The evidence shows that the final state of the water (in particular, its temperature and volume) is the same in every case. It is irrelevant if the work is electrical, mechanical, chemical,... or if done suddenly or slowly, as long as it is performed in an adiabatic way, that is to say, without heat transfer into or out of the system.

Evidence of this kind shows that to increase the temperature of the water in the tank, the qualitative kind of adiabatically performed work does not matter. No qualitative kind of adiabatic work has ever been observed to decrease the temperature of the water in the tank.

A change from one state to another, for example an increase of both temperature and volume, may be conducted in several stages, for example by externally supplied electrical work on a resistor in the body, and adiabatic expansion allowing the body to do work on the surroundings. It needs to be shown that the time order of the stages, and their relative magnitudes, does not affect the amount of adiabatic work that needs to be done for the change of state. According to one respected scholar: "Unfortunately, it does not seem that experiments of this kind have ever been carried out carefully. ... We must therefore admit that the statement which we have enunciated here, and which is equivalent to the first law of thermodynamics, is not well founded on direct experimental evidence." Another expression of this view is "... no systematic precise experiments to verify this generalization directly have ever been attempted."

This kind of evidence, of independence of sequence of stages, combined with the above-mentioned evidence, of independence of qualitative kind of work, would show the existence of an important state variable that corresponds with adiabatic work, but not that such a state variable represented a conserved quantity. For the latter, another step of evidence is needed, which may be related to the concept of reversibility, as mentioned below.

That important state variable was first recognized and denoted U by Clausius in 1850, but he did not then name it, and he defined it in terms not only of work but also of heat transfer in the same process. It was also independently recognized in 1850 by Rankine, who also denoted it U ; and in 1851 by Kelvin who then called it "mechanical energy", and later "intrinsic energy". In 1865, after some hestitation, Clausius began calling his state function U "energy". In 1882 it was named as the *internal energy* by Helmholtz. If only adiabatic processes were of interest, and heat could be ignored, the concept of internal energy would hardly arise or be needed. The relevant physics would be largely covered by the concept of potential energy, as was intended in the 1847 paper of Helmholtz on the principle of conservation of energy, though that did not deal with forces that cannot be described by a potential, and thus did not fully justify the principle. Moreover, that paper was critical of the early work of Joule that had by then been performed. A great merit of the internal energy concept is that it frees thermodynamics from a restriction to cyclic processes, and allows a treatment in terms of thermodynamic states.

In an adiabatic process, adiabatic work takes the system either from a reference state O with internal energy U(O) to an arbitrary one A with internal energy U(A), or from the state A to the state O:

$$U(A) = U(O) - W_{O \to A}^{\text{adiabatic}} \text{ or } U(O) = U(A) - W_{A \to O}^{\text{adiabatic}}.$$

Except under the special, and strictly speaking, fictional, condition of reversibility, only one of the processes adiabatic,$O \to A$ or adiabatic,$A \to O$ is empirically feasible by a simple application of externally supplied work. The reason for this is given as the second law of thermodynamics and is not considered in this article.

The fact of such irreversibility may be dealt with in two main ways, according to different points of view:

- Since the work of Bryan (1907), the most accepted way to deal with it nowadays, followed by Carathéodory, is to rely on the previously established concept of quasi-static processes, as follows. Actual physical processes of transfer of energy as work are always at least to some degree irreversible. The irreversibility is often due to mechanisms known as dissipative, that transform bulk kinetic energy into internal energy. Examples are friction and viscosity. If the process is performed more slowly, the frictional or viscous dissipation is less. In the limit of infinitely slow performance, the dissipation tends to zero and then the limiting process, though fictional rather than actual, is notionally reversible, and is called quasi-static. Throughout the course of the fictional limiting quasi-static process, the internal intensive variables of the system are equal to the external intensive variables, those that describe the reactive forces exerted by the surroundings. This can be taken to justify the formula

$$W_{A \to O}^{\text{adiabatic, quasi-static}} = -W_{O \to A}^{\text{adiabatic, quasi-static}}.$$

- Another way to deal with it is to allow that experiments with processes of heat transfer to or from the system may be used to justify the formula above. Moreover, it deals to some extent with the problem of lack of direct experimental evidence that the time order of stages of a process does not matter in the determination of internal energy. This way does not provide theoretical purity in terms of adiabatic work processes, but is empirically feasible, and is in accord with experiments actually done, such as the Joule experiments mentioned just above, and with older traditions.

The formula above allows that to go by processes of quasi-static adiabatic work from the state A to the state B we can take a path that goes through the reference state O, since the quasi-static adiabatic work is independent of the path

$$-W_{A \to B}^{\text{adiabatic, quasi-static}} = -W_{A \to O}^{\text{adiabatic, quasi-static}} - W_{O \to B}^{\text{adiabatic, quasi-static}}$$

$$= W^{\text{adiabatic, quasi-static}}_{O \to A} - W^{\text{adiabatic, quasi-static}}_{O \to B} = -U(A) + U(B) = \Delta U$$

This kind of empirical evidence, coupled with theory of this kind, largely justifies the following statement:

> *For all adiabatic processes between two specified states of a closed system of any nature, the net work done is the same regardless the details of the process, and determines a state function called internal energy, U ."*

Adynamic Processes

A complementary observable aspect of the first law is about heat transfer. Adynamic transfer of energy as heat can be measured empirically by changes in the surroundings of the system of interest by calorimetry. This again requires the existence of adiabatic enclosure of the entire process, system and surroundings, though the separating wall between the surroundings and the system is thermally conductive or radiatively permeable, not adiabatic. A calorimeter can rely on measurement of sensible heat, which requires the existence of thermometers and measurement of temperature change in bodies of known sensible heat capacity under specified conditions; or it can rely on the measurement of latent heat, through measurement of masses of material that change phase, at temperatures fixed by the occurrence of phase changes under specified conditions in bodies of known latent heat of phase change. The calorimeter can be calibrated by adiabatically doing externally determined work on it. The most accurate method is by passing an electric current from outside through a resistance inside the calorimeter. The calibration allows comparison of calorimetric measurement of quantity of heat transferred with quantity of energy transferred as work. According to one textbook, "The most common device for measuring ΔU is an adiabatic bomb calorimeter." According to another textbook, "Calorimetry is widely used in present day laboratories." According to one opinion, "Most thermodynamic data come from calorimetry..." According to another opinion, "The most common method of measuring "heat" is with a calorimeter."

When the system evolves with transfer of energy as heat, without energy being transferred as work, in an adynamic process, the heat transferred to the system is equal to the increase in its internal energy:

$$Q^{\text{adynamic}}_{A \to B} = \Delta U.$$

General Case for Reversible Processes

Heat transfer is practically reversible when it is driven by practically negligibly small temperature gradients. Work transfer is practically reversible when it occurs so slowly that there are no frictional effects within the system; frictional effects outside the system should also be zero if the process is to be globally reversible. For a particular reversible process in general, the work done reversibly on the system, $W_{A \to B}^{\text{path } P_0, \text{reversible}}$, and the heat transferred reversibly to the system, $Q_{A \to B}^{\text{path } P_0, \text{reversible}}$ are not required to occur respectively adiabatically or adynamically, but they must belong to the same particular process defined by its particular reversible path, P_0, through the space of thermodynamic states. Then the work and heat transfers can occur and be calculated simultaneously.

Putting the two complementary aspects together, the first law for a particular reversible process can be written

$$-W_{A \to B}^{\text{path } P_0, \text{reversible}} + Q_{A \to B}^{\text{path } P_0, \text{reversible}} = \Delta U.$$

This combined statement is the expression the first law of thermodynamics for reversible processes for closed systems.

In particular, if no work is done on a thermally isolated closed system we have

$$\Delta U = 0.$$

This is one aspect of the law of conservation of energy and can be stated:

The internal energy of an isolated system remains constant.

General Case for Irreversible Processes

If, in a process of change of state of a closed system, the energy transfer is not under a practically zero temperature gradient and practically frictionless, then the process is irreversible. Then the heat and work transfers may be difficult to calculate, and irreversible thermodynamics is called for. Nevertheless, the first law still holds and provides a check on the measurements and calculations of the work done irreversibly on the system, $W_{A \to B}^{\text{path } P_1, \text{irreversible}}$, and the heat transferred irreversibly to the system, $Q_{A \to B}^{\text{path } P_1, \text{irreversible}}$, which belong to the same particular process defined by its particular irreversible path, P_1, through the space of thermodynamic states.

$$-W_{A \to B}^{\text{path } P_1, \text{irreversible}} + Q_{A \to B}^{\text{path } P_1, \text{irreversible}} = \Delta U.$$

This means that the internal energy U is a function of state and that the internal energy change ΔU between two states is a function only of the two states.

Overview of the Weight of Evidence for the Law

The first law of thermodynamics is so general that its predictions cannot all be directly tested. In many properly conducted experiments it has been precisely supported, and never violated. Indeed, within its scope of applicability, the law is so reliably established, that, nowadays, rather than experiment being considered as testing the accuracy of the law, it is more practical and realistic to think of the law as testing the accuracy of experiment. An experimental result that seems to violate the law may be assumed to be inaccurate or wrongly conceived, for example due to failure to account for an important physical factor. Thus, some may regard it as a principle more abstract than a law.

State Functional Formulation for Infinitesimal Processes

When the heat and work transfers in the equations above are infinitesimal in magnitude, they are often denoted by δ, rather than exact differentials denoted by d, as a reminder that heat and work do not describe the *state* of any system. The integral of an inexact differential depends upon the particular path taken through the space of thermodynamic parameters while the integral of an exact differential depends only upon the initial and final states. If the initial and final states are the same, then the integral of an inexact differential may or may not be zero, but the integral of an exact differential is always zero. The path taken by a thermodynamic system through a chemical or physical change is known as a thermodynamic process.

The first law for a closed homogeneous system may be stated in terms that include concepts that are established in the second law. The internal energy U may then be expressed as a function of the system's defining state variables S, entropy, and V, volume: $U = U(S, V)$. In these terms, T, the system's temperature, and P, its pressure, are partial derivatives of U with respect to S and V. These variables are important throughout thermodynamics, though not necessary for the statement of the first law. Rigorously, they are defined only when the system is in its own state of internal thermodynamic equilibrium. For some purposes, the concepts provide good approximations for scenarios sufficiently near to the system's internal thermodynamic equilibrium.

The first law requires that:

$$dU = \delta Q - \delta W$$ (closed system, general process, quasi- static or irreversible).

Then, for the fictive case of a reversible process, dU can be written in terms of exact differentials. One may imagine reversible changes, such that there is at each instant negligible departure from thermodynamic equilibrium within the system. This excludes isochoric work. Then, mechanical work is given by $\delta W = -P\,dV$ and the quantity of heat added can be expressed as $\delta Q = T\,dS$. For these conditions

$$dU = TdS - PdV$$ (closed system, reversible process).

While this has been shown here for reversible changes, it is valid in general, as U can be considered as a thermodynamic state function of the defining state variables S and V:

$$dU = TdS - PdV$$ (closed system, general process, quasi-static or irreversible).

Above equation is known as the fundamental thermodynamic relation for a closed system in the energy representation, for which the defining state variables are S and V, with respect to which T and P are partial derivatives of U. It is only in the fictive reversible case, when isochoric work is excluded, that the work done and heat transferred are given by $-P\,dV$ and $T\,dS$.

In the case of a closed system in which the particles of the system are of different types and, because chemical reactions may occur, their respective numbers are not necessarily constant, the fundamental thermodynamic relation for dU becomes:

$$dU = TdS - PdV + \sum_i \mu_i dN_i.$$

where dN_i is the (small) increase in amount of type-i particles in the reaction, and μ_i is known as the chemical potential of the type-i particles in the system. If dN_i is expressed in mol then μ_i is expressed in J/mol. If the system has more external mechanical variables than just the volume that can change, the fundamental thermodynamic relation further generalizes to:

$$dU = TdS - \sum_i X_i dx_i + \sum_j \mu_j dN_j.$$

Here the X_i are the generalized forces corresponding to the external variables x_i. The parameters X_i are independent of the size of the system and are called intensive parameters and the x_i are proportional to the size and called extensive parameters.

For an open system, there can be transfers of particles as well as energy into or out of the system during a process. For this case, the first law of thermodynamics still holds, in the form that the internal energy is a function of state and the change of internal energy in a process is a function only of its initial and final states, as noted in the section below headed First law of thermodynamics for open systems.

A useful idea from mechanics is that the energy gained by a particle is equal to the force applied to the particle multiplied by the displacement of the particle while that force is applied. Now consider the first law without the heating term: $dU = -PdV$. The pressure P can be viewed as a force (and in fact has units of force per unit area) while dV is the displacement (with units of distance times area). We may say, with respect to this work term, that a pressure difference forces a transfer of volume, and that the product of the two (work) is the amount of energy transferred out of the system as a result of the process. If one were to make this term negative then this would be the work done on the system.

It is useful to view the TdS term in the same light: here the temperature is known as a "generalized" force (rather than an actual mechanical force) and the entropy is a generalized displacement.

Similarly, a difference in chemical potential between groups of particles in the system drives a chemical reaction that changes the numbers of particles, and the corresponding product is the amount of chemical potential energy transformed in process. For example, consider a system consisting of two phases: liquid water and water vapor. There is a generalized "force" of evaporation that drives water molecules out of the liquid. There is a generalized "force" of condensation that drives vapor molecules out of the vapor. Only when these two "forces" (or chemical potentials) are equal is there equilibrium, and the net rate of transfer zero.

The two thermodynamic parameters that form a generalized force-displacement pair are called "conjugate variables". The two most familiar pairs are, of course, pressure-volume, and temperature-entropy.

Spatially Inhomogeneous Systems

Classical thermodynamics is initially focused on closed homogeneous systems (e.g. Planck 1897/1903), which might be regarded as 'zero-dimensional' in the sense that they have no spatial variation. But it is desired to study also systems with distinct internal motion and spatial inhomogeneity. For such systems, the principle of conservation of energy is expressed in terms not only of internal energy as defined for homogeneous systems, but also in terms of kinetic energy and potential energies of parts of the inhomogeneous system with respect to each other and with respect to long-range external forces. How the total energy of a system is allocated between these three more specific kinds of energy varies according to the purposes of different writers; this is because these components of energy are to some extent mathematical artefacts rather than actually measured physical quantities. For any closed homogeneous component of an inhomogeneous closed system, if E denotes the total energy of that component system, one may write

$$E = E^{kin} + E^{pot} + U$$

where E^{kin} and E^{pot} denote respectively the total kinetic energy and the total potential energy of the component closed homogeneous system, and U denotes its internal energy.

Potential energy can be exchanged with the surroundings of the system when the surroundings impose a force field, such as gravitational or electromagnetic, on the system.

A compound system consisting of two interacting closed homogeneous component subsystems has a potential energy of interaction E_{12}^{pot} between the subsystems. Thus, in an obvious notation, one may write

$$E = E_1^{kin} + E_1^{pot} + U_1 + E_2^{kin} + E_2^{pot} + U_2 + E_{12}^{pot}$$

The quantity E_{12}^{pot} in general lacks an assignment to either subsystem in a way that is not arbitrary, and this stands in the way of a general non-arbitrary definition of transfer of energy as work. On occasions, authors make their various respective arbitrary assign-ments.

The distinction between internal and kinetic energy is hard to make in the presence of turbulent motion within the system, as friction gradually dissipates macroscopic ki-netic energy of localised bulk flow into molecular random motion of molecules that is classified as internal energy. The rate of dissipation by friction of kinetic energy of localised bulk flow into internal energy, whether in turbulent or in streamlined flow, is an important quantity in non-equilibrium thermodynamics. This is a serious difficulty for attempts to define entropy for time-varying spatially inhomogeneous systems.

First law of Thermodynamics for Open Systems

For the first law of thermodynamics, there is no trivial passage of physical conception from the closed system view to an open system view. For closed systems, the concepts of an adiabatic enclosure and of an adiabatic wall are fundamental. Matter and internal energy cannot permeate or penetrate such a wall. For an open system, there is a wall that allows penetration by matter. In general, matter in diffusive motion carries with it some internal energy, and some microscopic potential energy changes accompany the motion. An open system is not adiabatically enclosed.

There are some cases in which a process for an open system can, for particular purposes, be considered as if it were for a closed system. In an open system, by definition hypothet-ically or potentially, matter can pass between the system and its surroundings. But when, in a particular case, the process of interest involves only hypothetical or potential but no actual passage of matter, the process can be considered as if it were for a closed system.

Internal Energy for an Open System

Since the revised and more rigorous definition of the internal energy of a closed sys-tem rests upon the possibility of processes by which adiabatic work takes the system from one state to another, this leaves a problem for the definition of internal energy for an open system, for which adiabatic work is not in general possible. According to Max Born, the transfer of matter and energy across an open connection "cannot be re-duced to mechanics". In contrast to the case of closed systems, for open systems, in the presence of diffusion, there is no unconstrained and unconditional physical distinction between convective transfer of internal energy by bulk flow of matter, the transfer of internal energy without transfer of matter (usually called heat conduction and work transfer), and change of various potential energies. The older traditional way and the conceptually revised (Carathéodory) way agree that there is no physically unique defi-nition of heat and work transfer processes between open systems.

In particular, between two otherwise isolated open systems an adiabatic wall is by definition impossible. This problem is solved by recourse to the principle of conservation of energy. This principle allows a composite isolated system to be derived from two other component non-interacting isolated systems, in such a way that the total energy of the composite isolated system is equal to the sum of the total energies of the two component isolated systems. Two previously isolated systems can be subjected to the thermodynamic operation of placement between them of a wall permeable to matter and energy, followed by a time for establishment of a new thermodynamic state of internal equilibrium in the new single unpartitioned system. The internal energies of the initial two systems and of the final new system, considered respectively as closed systems as above, can be measured. Then the law of conservation of energy requires that

$$\Delta U_s + \Delta U_o = 0,$$

where ΔU_s and ΔU_o denote the changes in internal energy of the system and of its surroundings respectively. This is a statement of the first law of thermodynamics for a transfer between two otherwise isolated open systems, that fits well with the conceptually revised and rigorous statement of the law stated above.

For the thermodynamic operation of adding two systems with internal energies U_1 and U_2, to produce a new system with internal energy U, one may write $U = U_1 + U_2$; the reference states for U, U_1 and U_2 should be specified accordingly, maintaining also that the internal energy of a system be proportional to its mass, so that the internal energies are extensive variables.

There is a sense in which this kind of additivity expresses a fundamental postulate that goes beyond the simplest ideas of classical closed system thermodynamics; the extensivity of some variables is not obvious, and needs explicit expression; indeed one author goes so far as to say that it could be recognized as a fourth law of thermodynamics, though this is not repeated by other authors.

Also of course

$$\Delta N_s + \Delta N_o = 0,$$

where ΔN_s and ΔN_o denote the changes in mole number of a component substance of the system and of its surroundings respectively. This is a statement of the law of conservation of mass.

Process of Transfer of Matter Between an Open System and its Surroundings

A system connected to its surroundings only through contact by a single permeable wall, but otherwise isolated, is an open system. If it is initially in a state of contact equilibrium with a surrounding subsystem, a thermodynamic process of transfer of matter can be made to occur between them if the surrounding subsystem is subjected to some thermodynamic operation, for example, removal of a partition between it and some further surrounding subsystem. The removal of the partition in the surroundings initiates a process of exchange between the system and its contiguous surrounding subsystem.

An example is evaporation. One may consider an open system consisting of a collection of liquid, enclosed except where it is allowed to evaporate into or to receive condensate from its vapor above it, which may be considered as its contiguous surrounding subsystem, and subject to control of its volume and temperature.

A thermodynamic process might be initiated by a thermodynamic operation in the surroundings, that mechanically increases in the controlled volume of the vapor. Some mechanical work will be done within the surroundings by the vapor, but also some of the parent liquid will evaporate and enter the vapor collection which is the contiguous surrounding subsystem. Some internal energy will accompany the vapor that leaves the system, but it will not make sense to try to uniquely identify part of that internal energy as heat and part of it as work. Consequently, the energy transfer that accompanies the transfer of matter between the system and its surrounding subsystem cannot be uniquely split into heat and work transfers to or from the open system. The component of total energy transfer that accompanies the transfer of vapor into the surrounding subsystem is customarily called 'latent heat of evaporation', but this use of the word heat is a quirk of customary historical language, not in strict compliance with the thermodynamic definition of transfer of energy as heat. In this example, kinetic energy of bulk flow and potential energy with respect to long-range external forces such as gravity are both considered to be zero. The first law of thermodynamics refers to the change of internal energy of the open system, between its initial and final states of internal equilibrium.

Open System with Multiple Contacts

An open system can be in contact equilibrium with several other systems at once.

This includes cases in which there is contact equilibrium between the system, and several subsystems in its surroundings, including separate connections with subsystems through walls that are permeable to the transfer of matter and internal energy as heat and allowing friction of passage of the transferred matter, but immovable, and separate connections through adiabatic walls with others, and separate connections through diathermic walls impermeable to matter with yet others. Because there are physically separate connections that are permeable to energy but impermeable to matter, between the system and its surroundings, energy transfers between them can occur with definite heat and work characters. Conceptually essential here is that the internal energy transferred with the transfer of matter is measured by a variable that is mathematically independent of the variables that measure heat and work.

With such independence of variables, the total increase of internal energy in the process is then determined as the sum of the internal energy transferred from the surroundings with the transfer of matter through the walls that are permeable to it, and of the internal energy transferred to the system as heat through the diathermic walls, and of the energy transferred to the system as work through the adiabatic walls, including the energy transferred to the system by long-range forces. These simultaneously transferred quantities of energy are defined by events in the surroundings of the system. Because the internal energy transferred with matter is not in general uniquely resolvable into heat and work components, the total energy transfer cannot in general be uniquely resolved into heat and work components. Under these conditions, the following formula can describe the process in terms of externally defined thermodynamic variables, as a statement of the first law of thermodynamics:

$$\Delta U_0 = Q - W - \sum_{i=1}^{m} \Delta U_i$$

(suitably defined surrounding subsystems, general process, quasi-static or irreversible),

where ΔU_0 denotes the change of internal energy of the system, and ΔU_i denotes the change of internal energy of the ith of the m surrounding subsystems that are in open contact with the system, due to transfer between the system and that ith surrounding subsystem, and Q denotes the internal energy transferred as heat from the heat reservoir of the surroundings to the system, and W denotes the energy transferred from the system to the surrounding subsystems that are in adiabatic connection with it. The case of a wall that is permeable to matter and can move so as to allow transfer of energy as work is not considered here.

Combination of First and Second Laws

If the system is described by the energetic fundamental equation, $U_0 = U_0(S, V, N_j)$, and if the process can be described in the quasi-static formalism, in terms of the internal state variables of the system, then the process can also be described by a combination of the first and second laws of thermodynamics, by the formula

$$dU_0 = TdS - PdV + \sum_{j=1}^{n} \mu_j dN_j$$

where there are n chemical constituents of the system and permeably connected surrounding subsystems, and where T, S, P, V, N_j, and μ_j, are defined as above.

For a general natural process, there is no immediate term-wise correspondence between above given equations, because they describe the process in different conceptual frames.

Nevertheless, a conditional correspondence exists. There are three relevant kinds of wall here: purely diathermal, adiabatic, and permeable to matter. If two of those kinds of wall are sealed off, leaving only one that permits transfers of energy, as work, as heat, or with matter, then the remaining permitted terms correspond precisely. If two of the kinds of wall are left unsealed, then energy transfer can be shared between them, so that the two remaining permitted terms do not correspond precisely.

For the special fictive case of quasi-static transfers, there is a simple correspondence. For this, it is supposed that the system has multiple areas of contact with its surroundings. There are pistons that allow adiabatic work, purely diathermal walls, and open connections with surrounding subsystems of completely controllable chemical potential (or equivalent controls for charged species). Then, for a suitable fictive quasi-static transfer, one can write

$$\delta Q = T dS \quad \text{and} \quad \delta W = P dV$$

(suitably defined surrounding subsystems, quasi-static transfers of energy).

For fictive quasi-static transfers for which the chemical potentials in the connected surrounding subsystems are suitably controlled, these can be put into above equation to yield

$$dU_0 = \delta Q - \delta W + \sum_{j=1}^{n} \mu_j dN_j$$

(suitably defined surrounding subsystems, quasi-static transfers).

The reference does not actually write above equation, but what it does write is fully compatible with it. Another helpful account is given by Tschoegl.

There are several other accounts of this, in apparent mutual conflict.

Non-equilibrium Transfers

The transfer of energy between an open system and a single contiguous subsystem of its surroundings is considered also in non-equilibrium thermodynamics. The problem of definition arises also in this case. It may be allowed that the wall between the system and the subsystem is not only permeable to matter and to internal energy, but also may be movable so as to allow work to be done when the two systems have different pressures. In this case, the transfer of energy as heat is not defined.

Methods for study of non-equilibrium processes mostly deal with spatially continuous flow systems. In this case, the open connection between system and surroundings is usually taken to fully surround the system, so that there are no separate connections impermeable to matter but permeable to heat. Except for the special case mentioned above when there is no actual transfer of matter, which can be treated as if for a closed system, in strictly defined thermodynamic terms, it follows that transfer of energy as heat is not defined. In this sense, there is no such thing as 'heat flow' for a continuous-flow open system. Properly, for closed systems, one speaks of transfer of internal energy as heat, but in general, for open systems, one can speak safely only of transfer of internal energy. A factor here is that there are often cross-effects between distinct transfers, for example that transfer of one substance may cause transfer of another even when the latter has zero chemical potential gradient.

Usually transfer between a system and its surroundings applies to transfer of a state variable, and obeys a balance law, that the amount lost by the donor system is equal to the amount gained by the receptor system. Heat is not a state variable. For his 1947 definition of "heat transfer" for discrete open systems, the author Prigogine carefully explains at some length that his definition of it does not obey a balance law. He describes this as paradoxical.

The situation is clarified by Gyarmati, who shows that his definition of "heat transfer", for continuous-flow systems, really refers not specifically to heat, but rather to transfer of internal energy, as follows. He considers a conceptual small cell in a situation of continuous-flow as a system defined in the so-called Lagrangian way, moving with the local center of mass. The flow of matter across the boundary is zero when considered as a flow of total mass. Nevertheless, if the material constitution is of several chemically distinct components that can diffuse with respect to one another, the system is considered to be open, the diffusive flows of the components being defined with respect to the center of mass of the system, and balancing one another as to mass transfer. Still there can be a distinction between bulk flow of internal energy and diffusive flow of internal energy in this case, because the internal energy density does not have to be constant per unit mass of material, and allowing for non-conservation of internal energy because of local conversion of kinetic energy of bulk flow to internal energy by viscosity.

Gyarmati shows that his definition of "the heat flow vector" is strictly speaking a definition of flow of internal energy, not specifically of heat, and so it turns out that his use here of the word heat is contrary to the strict thermodynamic definition of heat, though it is more or less compatible with historical custom, that often enough did not clearly distinguish between heat and internal energy; he writes "that this relation must be considered to be the exact definition of the concept of heat flow, fairly loosely used in experimental physics and heat technics." Apparently in a different frame of thinking from that of the above-mentioned paradoxical usage in the earlier sections of the historic 1947 work by Prigogine, about discrete systems, this usage of Gyarmati is consistent with the later sections of the same 1947 work by Prigogine, about continuous-flow systems, which use the term "heat flux" in just this way. This usage is also followed by Glansdorff and Prigogine in their 1971 text about continuous-flow systems. They write: "Again the flow of internal energy may be split into a convection flow $\rho u v$ and a conduction flow. This conduction flow is by definition the heat flow W. Therefore: $j[U] = \rho u v + W$ where u denotes the [internal] energy per unit mass. [These authors actually use the symbols E and e to denote internal energy but their notation has been changed here to accord with the notation of this article. These authors actually use the symbol *U* to refer to total energy, including kinetic energy of bulk flow.]" This usage is followed also by other writers on non-equilibrium thermodynamics such as Lebon, Jou, and Casas-Vásquez, and de Groot and Mazur. This usage is described by Bailyn as stating the non-convective flow of internal energy, and is listed as his definition number 1, according to the first law of thermodynamics. This usage is also followed by workers in the kinetic theory of gases. This is not the *ad hoc* definition of "reduced heat flux" of Haase.

In the case of a flowing system of only one chemical constituent, in the Lagrangian representation, there is no distinction between bulk flow and diffusion of matter. Moreover, the flow of matter is zero into or out of the cell that moves with the local center of mass. In effect, in this description, one is dealing with a system effectively closed to the transfer of matter. But still one can validly talk of a distinction between bulk flow and diffusive flow of internal energy, the latter driven by a temperature gradient within the flowing material, and being defined with respect to the local center of mass of the bulk flow. In this case of a virtually closed system, because of the zero matter transfer, as noted above, one can safely distinguish between transfer of energy as work, and transfer of internal energy as heat.

Various Heat Effects

1. Effect of temperature

The variation of heat content in pure compounds with temperature can be determined by

$$H_T - H_{298} = aT + bT^2 + cT^{-1} + d$$

$H_T - H_{298}$ is increase in heat content in cal/mole as the substance is heated from 298K to T. In all calculations the reference temperature is 298K (25°C or 77°F) T is temperature in K and a, b, c and d are constants and depends on state or aggregation of the system. The quantity $H_T - H_{298}$ is called sensible heat, `

Molal heat capacity at constant pressure is

$$C_p = a + 2bT - CT^{-2}$$

$$C_p = \left(\frac{\partial H}{\partial t}\right)_p \quad \therefore H_2 - H_1 = \int_{T_1}^{T_2} C_p dT$$

Also change in heat content can be determined by mean heat capacity C_m

$$H_2 - H_1 = C_m(T_2 - T_1)$$

$$C_m = \frac{\int_{T_1}^{T_2} C_p dT}{T_2 - T_1}$$

We may write

$$H_2^l - H_1^l = C_m(T_2 - T_1)$$

Equation may be used to calculate heat content for rapid calculations. More accurate would be to use above equations.

1.   Changes in state of aggregation

When solid is heated to melting point additional heat must be supplied to melt it. At the melting print, this additional heat does not increase the temperature. The heat content for melting at constant pressure is called latent heat of fusion (ΔH_{fusion}) and this heat must be supplied at the melting point to transform solid into liquid. An equal quantity of heat is liberated during solidification so that

$$(\Delta H_{solidification}) = -(\Delta H_{fusion})$$

Negative sign indicates liberation of heat. Similarly, heat effects accompanying evaporation and allotropic changes in solids are measured by latent heat of vaporization and heats of transformation.

Value of ΔH for changes in state of aggregation vary with temperature and pressure under which change is carried out. For example heat of vaporization of water at 1000C is 542cal/g, whereas it is 583 cal/g at 25°C.

Example

Calculate latent heat of fusion of Cu

The melting point of Cu is 1357K.

For solid $Cu : H_T - H_{298} = 5.41T + 0.75 \times 10^{-3}T^2 - 1680$

$= 7042\ cal/mol$ at melting point

For liquid $C_u H_T - H_{298} = 7.50T - 20$

$= 10158\ cal/mol$ atmelting point

Therefore $\Delta H_{fusion} = 10158 - 7042 = 3116 cal/mol$

Heat of Formation

The formation of chemical compound from its elements is associated with either absorption or liberation of heat. Thus the formation of $H_2o(l)$ form the elements at 298K (25 C) can be written as

$$H_2(g) + \frac{1}{2}O_2(g) = H_2O(l)$$

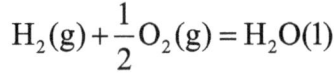

$$\Delta H^0_{298} = -68320 \text{ cal / mol at 1 atm pressure}$$

Heat of Reaction

ΔH for any process depends only on the initial and final states and not on the path. If a process is divided into several steps, and ΔH is determined for each step, the algebraic sum of the ΔH values of all steps will be equal to ΔH for the original process. Consider an example to calculate heat of reaction of the following reaction:

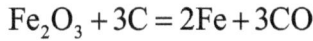

$$Fe_2O_3 + 3C = 2Fe + 3CO$$

This Reaction Comprises of the Following Reactions:

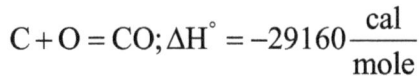

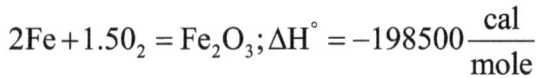

$$C + O = CO; \Delta H^\circ = -29160 \frac{\text{cal}}{\text{mole}}$$

$$2Fe + 1.5O_2 = Fe_2O_3; \Delta H^\circ = -198500 \frac{\text{cal}}{\text{mole}}$$

Heat of formation for $Fe_2O_3 + 3C = 2Fe + 3CO$ reaction at 298K is

$$(\Delta H^\circ_f)_{298} = 3\Delta H^\circ_{co} - \Delta H^\circ_{Fe_2O_3}$$

It must be noted that heat of formation of element is zero by using the value of ΔH^0 we get $(\Delta H^\circ_f)_{298} = 111020 \text{cal / molFe}_2O_3$ (endothermic), that means $Fe_2O_3 + 3C = 2Fe + 3CO$ reaction is accompanied by absorption of heat.

Consider the Reaction

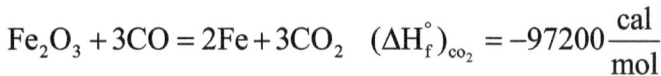

$$Fe_2O_3 + 3CO = 2Fe + 3CO_2 \quad (\Delta H^\circ_f)_{co_2} = -97200 \frac{\text{cal}}{\text{mol}}$$

$$(\Delta H^\circ_f)_{298} = 3 \times (\Delta H^\circ_f)_{co_2} - (\Delta H^\circ_f)_{Fe_2O_3} - 3(\Delta H^\circ_f)CO$$

$= -5620 \text{cal / molFe}_2O_3$. This is exothermic reaction.

The heat of reaction calculated above are at 298K (25°C) Heat of reaction depends on temperature and can be determined

$$(\Delta H^\circ_f)_{T_2} = (\Delta H_f)_{298} + \Sigma(H_{T2} - H_{298})_{products} - \Sigma(H_{T2} - H_{T1})_{reac \tan ts}$$

Material Balance

The material and heat balance for an operating process is an essential and routine plant record, in much the same way as income and expenditure record is a part of operating a business.

The material balance shows the weights and analysis of input and output materials and the calculated inputs and outputs of each of the important metal elements and compounds.

This accounting also serves as a check on plant data in that the various totals of inputs and outputs should be equal.

The material balance is the starting point for other calculations like energy requirements and cost calculations. From the energy calculations, one can design the energy flow path and identify energy losses to optimize energy consumption.

Basics of Materials Balance

Law of conservation of Mass

Mass of isolated systems remains constants irrespective of the changes occurring within in the system. Consider an open system of constant volume & mass.

Mass is transferred in to and out of the system; mass will accumulate if transfer rates are unequal Material balance at unsteady state is

$$\begin{bmatrix} Rate\ of\ Mass \\ in \end{bmatrix} = \begin{bmatrix} Rate\ of\ Mass \\ out \end{bmatrix} + \begin{bmatrix} Rate\ of\ accumlation \\ of\ mass \end{bmatrix}$$

$$\frac{dm}{dt} = \dot{m}_{in} - \dot{m}_{out}$$

Material balance equation is written for each component involved in the process;

$$\frac{dm_1}{dt} = (\dot{m}_{in})_1 - (\dot{m}_{out})_1$$

$$\frac{dm_2}{dt} = (\dot{m}_{in})_2 - (\dot{m}_{out})_2$$

$$\frac{dm_3}{dt} = (\dot{m}_{in})_3 - (\dot{m}_{out})_3$$

$$\frac{dm_i}{dt} = (\dot{m}_{in})_i - (\dot{m}_{out})_i$$

For a systems with multiple input and output streams, the material balance equation for Ith component

$$\frac{dm_i}{dt} = (\dot{m}_{in})_i - (\dot{m}_{out})_i$$

For steady state

$$\Sigma(\dot{m}_{in})_i - \Sigma(\dot{m}_{out})_i$$

For chemically reacting systems in addition to the law of conservation of mass, following two laws must also hold:

1. Law of definite proportions:

A given chemical compound always contains the same constitute elements in the same weight proportions. True for stoichiometric compounds.

2. Law of multiple proportions:

If two elements can form more than one compound, then the respective weights of one element that combines with a given weight of the other are in the ratio small whole numbers.

Basics of Energy Balance

Energy balance at steady state is

Heat input=Heat output

All sources of heat input should be considered for example

- Heat of reaction at 298 K

- Heat of fusion and evaporation

- Sensible heat in reactants

Many a time specific heat capacity is required to calculate sensible heat. Also heat of formation of compound would be required.

Similarly all different sources of heat output must also be considered. To mention few:

- Sensible heat in solid products

- Sensible heat in gases exiting the reactor

- Sensible heat in liquid products

- Heat losses

In all heat calculations a convenient reference temperature 298K is selected.

Metal Extraction Principles

Pyrometallurgical extraction of metals from natural oxide reserves involve the following steps:

- Removal of oxygen from the valuable mineral.

- Separation of metal from the gangue minerals.

For sulphide ore an additional step to convert sulphide into oxide is required. Several unit processes are employed to extract metal from ore by pyrometallurgical route. These are: roasting, matte smelting, reduction smelting, converting and refining. Different unit processes are combined to extract metal for example.

Production of metal	Combination of unit processes
Copper	Roasting +matte smelting +converting +refining Flash smelting +converting +refining.
Zinc & lead	Roasting +reduction smelting + refining
Nickel	Roasting +matte smelting + reduction smelting +refining.
Pig iron	Reduction smelting
Steel	Reduction smelting +refining.
Hydrometallurgical extraction is suitable for lean ores .here leaching followed by separation of metal from solution are the principal unit processes. Some examples of Hydrometallurgical extraction are	
Copper:	Roasting +leaching +cementation+ electro wining
Zinc:	Roasting +leaching +cementation+ electrolysis
Aluminum:	Leaching + electrolysis
Magnesium:	Leaching + electrolysis
Titanium:	Reduction smelting +leaching +chlorination +reduction by Mg

Mineral Beneficiation

Mineral beneficiation is the first step in extraction of metal from natural resources. With the depletion of high grade metal ores it is important to increase the metal grade of an ore by physical methods; which are termed mineral beneficiation. The objectives of mineral beneficiation are

- To increase the metal grade of ore

- To reduce the amount of gangue minerals so that lower volume of slag forms in pyromettallurgical extraction of metals. Slag contains mostly gangue minerals.

- To decrease the thermal energy required to separate liquid metal from gangue minerals.

- To decrease the aqueous solution requirement in hydrometturgical extraction of metals.

In this chapter mineral beneficiation science and technology are briefly reviewed so that readers can apply materials balance principles.

What Constitutes Mineral Beneficiation?

Ore is an aggregate of minerals and contains valuable and gangue minerals. The mineral beneficiation involves separations of gangue minerals from ore and is done in the following two stages:

1) Liberation of valuable mineral by size reduction technologies. In most ores the valuable minerals is distributed in the matrix of ore.

2) Concentration technologies to separate the gangue minerals and to achieve increase in the content of the valuable mineral to increase the metal grade.

Sizes Reduction Technologies

Size reduction or communication is an important step and may be used

- To produce particles of required sizes and shapes

- To liberate valuable mineral so that it can be concentrated.

- To increase the surface area available for chemical reaction.

It is often said that the efficiency of energy utilization during fragmentation of solid particles is only about 1% with respect to the new surface created. Energy consumption represents major cost in the mineral processing operation.

Crushing and grinding are size reduction methods. Crushing is applied to subsequent size reduction down to about 25mm. In grinding finer size is produced. Grinding or milling is an important size reduction method. In grinding force is applied by a medium which could either balls or rods .Both dry and wet grinding is done Wet grinding has the following characteristics.

- It requires less power

- It does not need dust central equipment.

- Wet grinding uses more steel grinding media to mill the material/per ton of product, as a result there occurs increase in erosion of the lining material.

- Water is required for wet grinding.

Material balance is important to determine

- Amount of water in a milling circuit

- % solid in slurry (slurry is a mixture of solid in water)

Both information are needed to determine the pump capacity to transport slurry. In wet milling water/solid ratio is important to control the viscosity of slurry. Too dilute slurry will lead to excessive wear of the medium .Too high a solid concentration results in cushioning of the medium. Percent solid in slurry can be determined by

$$\%\text{solid in slurry} = \frac{100\rho_s(\rho_m - 100)}{\rho_m(\rho_s - 1000)}$$

ρ_s, ρ_m is density of solid and slurry respectively. For example if ρ_s, is 3000 kg / m^3 $\rho_m = 1500 \text{kg / m}^3$ and then slurry contains 33.3% solid according to above equation.

Wet milling gives the following advantages:

- Less power requirement

- Required pollution as compared with dry milling.

After wet milling the milled product is classified by a hydro cyclone in to undersize (also termed underflow) and oversize (overflow) . The overflow is taken to the plant for concentration operation, for example flotation The undersize is recirculated after milling in a ball mill for further separation.

Concentration Technologies: Basics

The objectives of concentration technologies is to separate the valuable mineral from the gangue minerals .In all concentration methods feed is divided in three streams, namely concentrate, middling and tailings. Middlings are recycled within the plant and as such the plant output is two products, namely concentrate and tailings. Tailings are disposed whereas concentrate is sent to metal extraction.

Recovery and Grade

Recovery of the mineral in the concentrate and metal grade of the concentrate are important. Recovery is defined as

$$\text{Recovery} = 100 \times \frac{\text{Amount of valuable mineral in concentrate}}{\text{amount of feed}}$$

Grade of the concentrate can be defined either mineral grade or metal grade. Since concentrate is employed for metal extraction, metal grade is important (Note that concentrate contains mineral but not metal).

$$\text{Metal grade} = \frac{\text{Amount of metal in concentrate}}{\text{amount of concentrate}} \times 100$$

Metal grade means grade of valuable metal of the mineral in the concentrate. For example in the concentrate of chalcopyrite the grade of Cu is important. Similarly in the concentrate of galena, the grade of Pb is important. It must be clearly understood that ore does not contain metal. Metal grade is used to give an idea about the removal of gangue minerals and removal of oxygen or sulphur. For example mineral grade of pure Fe_2O_3 is unity but metal grade (or iron grade) of pure mineral Fe_2O_3 is 70% which means 30% oxygen has to be removed to get iron.

Recovery of a mineral in the concentrate can be 100% if all the feed is diverted in to concentrate. But metal grade will be very low. The maximum metal grade of the concentrate can be that of corresponding pure mineral, for example Cu grade in pure $CuFeS_2$ is 34.1%, lead grade in pure Pbs is 86.6% Zn grade in pure ZnS is 67%. Consider 1000 kg feed of chalcopyrite which produces 1000 kg concentrate. The concentrate contains $500kg\ CuFeS_2, 200kg\ Fe_2O_3, 200kg\ SiO_2$ and 10 kg Al_2O_3. The analysis of feed is same as that of concentrate.

Now the recovery $Cu\ Fe\ S_2$ in concentrate is 100%. But Cu grade in the

$$\text{concentrate} = \frac{500 \times 0.347}{1000} \times 100 = 17\%$$

Separation Efficiency

The feed of any concentration method is the milled product. The milled product is a mixture of particles of different Sizes, shapes and with different proportion of valuable mineral and gangue minerals. We have to separate particles containing valuable minerals for the maximum recovery. At the sametime metal grade should also be maximum of the concentrate.

In the following the valuable mineral in the concentrate is defined as metallic value of the valuable mineral.

Thus a concentrate is composed of metal in the valuable mineral +gangue. Let m is the metal grade of pure mineral and c is the metal grade of the concentrate.

$$\text{Gangue in concentrate} = (m-c)$$

$$\text{Gangue in feed} = (m-f)$$

f is the metal grade of the feed .If m_F is mass of feed and m_c is mass of concentrate

Recovery of gangue in concentrate $(R_g) = \dfrac{M_c(m-c)}{M_F(m-t)} \times 100$

Recovery of gangue in concentrate $(R_M) = \dfrac{M_c \times c}{M_F \times f} \times 100$

(SE) Separation efficiency $(R_M - R_g)100 = 100 \dfrac{M_c}{M_f}\left[\dfrac{m(c-f)}{f(m-t)}\right]$

Illustration on Separation Efficiency

Concentrate	Tin Grade	Recovery
1	63%	62%
2	42%	72%
3	21%	78%

Tin grade in feed is 1% and tin grade of pure mineral is 78.76% calculate separation efficiency (SE)

$$SE = 100 \frac{Mc}{MF}\left[\frac{m(c-f)}{f(m-f)}\right]$$

$$\frac{Mc}{MF} = \frac{f\,R_M}{c \times 100}$$

$$SE = \frac{f\,R_M}{c}\left[\frac{m(c-f)}{f(m-f)}\right]$$

Concentrate	Separation efficiency
1	61.8%
2	71.18%
3	75.24%

Note separation efficiency in concentrate 3 is high but tin grade is very low as compared to concentrate 1. But tin grade of concentrate is very high.

Concentration Methods

The most important processes are

1. Gravity concentration

2. Flotation

3. Magnetic and electrostatic separation

Gravity separation separates the minerals according to their different densities. It is used for the concentration of very heavy or very light minerals within a wide range of grain sizes.

In heavy media separation the density of pulp is intermediate between that of valuable mineral and gangue minerals. In that case light minerals float on top and the heavy minerals sink to the bottom of the pulp independent of particle size.

Other methods of gravity concentration utilize a combination of gravitational, inertial, frictional and viscous effects. Commonly used methods are jigging, washing tables, spirals etc.

Separation by flotation is based on the ability or lack of ability of different surfaces to be wetted by water. Hydrophobic minerals will cling to the air bubbles and rise with them, whereas the hydrophilic minerals will sink. Reagents like frothers, collectors, activators, depressors and conditioners are added to make the separation. Floatation has found its greatest application in the concentration of the sulphide minerals.

Electro statics and magnetic separation is based on differences in electrical conductivity of the mineral and magnetic properties the minerals respectively.

Gravity Separation

Gravity separation is an industrial method of separating two components, either a suspension, or dry granular mixture where separating the components with gravity is sufficiently practical: i.e. the components of the mixture have different specific weight. All of the gravitational methods are common in the sense that they all use gravity as the dominant force. The working principle is to lift the material by vacuum over an inclined vibrating screen covered deck. This results in the material being suspended in air while the heavier impurities are left behind on the screen and are discharged from the stone outlet. Gravity separation is used in a wide variety of industries, and can be most simply differentiated by the characteristics of the mixture to be separated - principally that of 'wet' i.e. - a suspension versus 'dry' -a mixture of granular product. Often other methods are applied to make the separation faster and more efficient, such as flocculation, coagulation and suction. The most notable advantages of the gravitational methods are their cost effectiveness and in some cases excellent reduction. Gravity separation is an attractive unit operation as it generally has low capital and operating costs, uses few if any chemicals that might cause environmental concerns and the recent development of new equipment enhances the range of separations possible.

Examples of Application

Agriculture: Gravity Separation tables are used for the removal of impurities, admixture, insect damage and immature kernels from the following examples: Wheat, Barley, Oilseed Rape, Peas, Beans, Cocoa Beans, Linseed. They can be used to separate and standardize coffee beans, cocoa beans, peanuts, corn, peas, rice, wheat, sesame and other food grains.

The Gravity Separator separates products of same size but with difference in specific weight. It has a vibrating rectangular deck, which makes it easy for the product to travel a longer distance, ensuring improved quality of the end product. The pressurized air in the deck enables the material to split according to its specific weight. As a result, the heavier particles travel to the higher level while the lighter particles travel to the lower level of the deck. It comes with easily adjustable air fans to control the volume of air distribution at different areas of the vibrating deck to meet the air supply needs of the deck. The table inclination, speed of eccentric motion and the feed rate can be precisely adjusted to achieve smooth operation of the machine.

Recycling: Gravity Separators are used to remove viable or valuable components from the recycling mixture i.e.: metal from plastic, rubber from plastic, different grades of plastic and these valuable materials are further recycled and reutilized

Preferential Flotation

Heavy liquids such as hhh tetrabromoethane can be used to separate ores from supporting rocks by preferential flotation. The rocks are crushed, and while sand, limestone, dolomite, and other types of rock material will float on TBE, ores such as sphalerite, galena and pyrite will sink.

Clarification/Thickening

Clarification is a name for the method of separating fluid from solid particles. Often clarification is used along with flocculation to make the solid particles sink faster to the bottom of the clarification pool while fluid is obtained from the surface which is free of solid particles.

Thickening is the same as clarification except reverse. Solids that sink to the bottom are obtained and fluid is rejected from the surface.

The difference of these methods could be demonstrated with the methods used in waste water processing: in the clarification phase, sludge sinks to the bottom of the pool and clear water flows over the clear water grooves and continues its journey. The obtained sludge is then pumped into the thickeners, where sludge thickens farther and is then obtained to be pumped into digestion to be prepared into fertilizer.

Sinking Chamber

When clearing gases, an often used and mostly working method for clearing large particles is to blow it into a large chamber where the gas's velocity reduces and the solid particles start sinking to the bottom. This method is used mostly because of its cheap cost.

Froth Flotation

Diagram of a cylindrical froth flotation cell with camera and light used in image analysis of the froth surface.

Froth flotation is a process for selectively separating hydrophobic materials from hydrophilic. This is used in mineral processing, paper recycling and waste-water treatment industries. Historically this was first used in the mining industry, where it was one of the great enabling technologies of the 20th century. It has been described as "the single most important operation used for the recovery and upgrading of sulfide ores". The development of froth flotation has improved the recovery of valuable minerals, such as copper- and lead-bearing minerals. Along with mechanized mining, it has allowed the economic recovery of valuable metals from much lower grade ore than previously.

History

Initially, naturally occurring chemicals such as fatty acids and oils were used as flotation reagents in a large quantity to increase the hydrophobicity of the valuable minerals. Since then, the process has been adapted and applied to a wide variety of materials to be separated, and additional collector agents, including surfactants and synthetic compounds have been adopted for various applications.

William Haynes patented a process in 1869 for separating sulfide and gangue minerals using oil and called it *bulk-oil flotation*. In 1885 Carrie Everson expanded upon this and patented a process calling for oil[s] but also an acid or a salt.

The first successful commercial flotation process for mineral sulphides was invented by Frank Elmore who worked on the development with his brother, Stanley. The Glasdir copper mine at Llanelltyd, near Dolgellau in North Wales was bought in 1896 by the Elmore brothers in conjunction with their father, William. In 1897, the Elmore brothers installed the world's first industrial size commercial flotation process for mineral beneficiation at the Glasdir mine. The process was not froth flotation but used oil to agglomerate (make balls of) pulverised sulphides and buoy them to the surface, and was patented in 1898 with a description of the process published in the *Engineering and Mining Journal*. By this time they had recognized the importance of air bubbles in assisting the oil to carry away the mineral particles. The Elmores had formed a company known as the Ore Concentration Syndicate Ltd to promote the commercial use of the process worldwide. However developments elsewhere, particularly in Australia by Minerals Separation Ltd., led to decades of hard fought legal battles and litigations which, ultimately, were lost as the process was superseded by more advanced techniques. Charles Butters, beginning about 1899, and working with both the Elmores and Minerals Separation's representative E.H. Nutter developed what was known to contemporaries as the "Butters Process". The flotation process was independently invented in the early 1900s in Australia by Charles Vincent Potter and around the same time by Guillaume Daniel Delprat. This process (developed circa 1902) did not use oil, but relied upon flotation by the generation of gas formed by the introduction of acid into the pulp. In 1902, Froment combined oil and gaseous flotation using a modification of the Potter-Delprat process.

Another process was developed in 1902 by Cattermole, who emulsified the pulp with a small quantity of oil, subjected it to violent agitation, then slow stirring which coagulated the target minerals into nodules which were separated from the pulp by gravity. This was the basis of the Minerals Separation Ltd. process. By 1904, the MacQuisten process (a surface tension based method) was developed but this would not work when slimes were present. In 1912, James M. Hyde modified the Minerals Separation Process and installed it in the Butte and Superior Mill in Basin, Montana, the first such installation in the USA.

John M. Callow, of General Engineering of Salt Lake City, had followed flotation from technical papers and the introduction in both the Butte and Superior Mill, and at Inspiration Copper in Arizona and determined that mechanical agitation was a drawback to the existing technology. Introducing a porous brick with compressed air, and a mechanical stirring mechanism, Callow applied for a patent in 1914. This method, known as Pneumatic Flotation, was recognized to revolutionize the process of flotation concentration. The American Institute of Mining Engineers presented Callow the James Douglas Gold Medal in 1926 for his contributions to the field of flotation.

In the 1960s the froth flotation technique was adapted for deinking recycled paper.

Industries

Mining

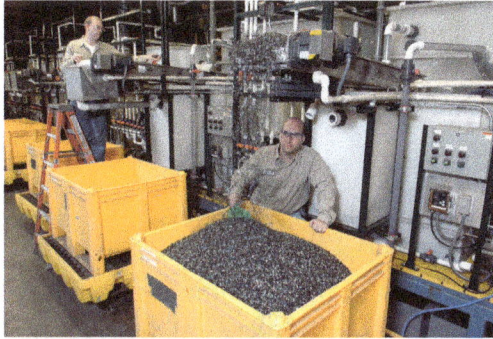

Froth flotation to separate plastics, Argonne National Laboratory

Froth flotation cells to concentrate copper and nickel sulfide minerals, Falconbridge, Ontario.

Froth flotation is a process for separating minerals from gangue by taking advantage of differences in their hydrophobicity. Hydrophobicity differences between valuable minerals and waste gangue are increased through the use of surfactants and wetting agents. The selective separation of the minerals makes processing complex (that is, mixed) ores economically feasible. The flotation process is used for the separation of a large range of sulfides, carbonates and oxides prior to further refinement. Phosphates and coal are also upgraded (purified) by flotation technology.

Prior to 1907, nearly all the copper mined in the US came from underground vein deposits, averaging 2.5 percent copper. By 1991, the average grade of copper ore mined in the US had fallen to only 0.6 percent.

Waste Water Treatment

The flotation process is also widely used in industrial waste water treatment plants, where it removes fats, oil, grease and suspended solids from waste water. These units are called dissolved air flotation (DAF) units. In particular, dissolved air flotation units are used in removing oil from the wastewater effluents of oil refineries, petrochemical and chemical plants, natural gas processing plants and similar industrial facilities.

Paper Recycling

Froth flotation is one of the processes used to recover recycled paper. In the paper industry this step is called deinking or just flotation. The target is to release and remove the hydrophobic contaminants from the recycled paper. The contaminants are mostly printing ink and stickies. Normally the setup is a two-stage system with 3,4 or 5 flotation cells in series.

Principle of Operation

Before froth flotation can work, the ore to be treated is reduced to fine particles by crushing and grinding (a process known as comminution) so that the various minerals exist as physically separate grains. This process is known as *liberation*. The particle sizes are typically less than 0.1 mm (100 µm), but sometimes sizes smaller than 7–10 µm are required. There is a tendency for the liberation size of the minerals to decrease over time as the ore bodies with coarse mineral grains that can be separated at larger sizes are depleted and replaced by ore bodies that were formerly considered too difficult.

In the mining industry, the plants where flotation is undertaken to concentrate ore are generally known as *concentrators* or *mills*.

For froth flotation, the ground ore is mixed with water to form a slurry and the desired mineral is rendered hydrophobic by the addition of a surfactant or *collector* chemical (although some mineral surfaces are naturally hydrophobic, requiring little or no addition of collector). The particular chemical depends on the nature of the mineral to be recovered and, perhaps, the natures of those that are not wanted. As an example, sodium ethyl xanthate may be added as a collector in the selective flotation of galena (lead sulfide) to separate it from sphalerite (zinc sulfide). This slurry (more properly called the *pulp*) of hydrophobic particles and hydrophilic particles is then introduced to tanks known as *flotation cells* that are aerated to produce bubbles. The hydrophobic particles attach to the air bubbles, which rise to the surface, forming a froth. The froth is removed from the cell, producing a concentrate ("con") of the target mineral.

Frothing agents, known as *frothers*, may be introduced to the pulp to promote the formation of a stable froth on top of the flotation cell.

The minerals that do not float into the froth are referred to as the *flotation tailings* or *flotation tails*. These tailings may also be subjected to further stages of flotation to recover the valuable particles that did not float the first time. This is known as *scavenging*. The final tailings after scavenging are normally pumped for disposal as mine fill or to tailings disposal facilities for long-term storage.

Froth flotation efficiency is determined by a series of probabilities: those of particle–bubble contact, particle–bubble attachment, transport between the pulp and the froth, and froth collection into the product launder. In a conventional mechanically-agitated cell, the void fraction (i.e. volume occupied by air bubbles) is low (5 to 10 percent) and the bubble size is usually greater than 1 mm. This results in a relatively low interfacial area and a low probability of particle–bubble contact. Consequently, several cells in series are required to increase the particle residence time, thus increasing the probability of particle–bubble contact.

Flotation is normally undertaken in several stages to maximize the recovery of the target mineral or minerals and the concentration of those minerals in the concentrate, while minimizing the energy input.

Flotation Stages

Roughing

The first stage is called *roughing*, which produces a *rougher concentrate*. The objective is to remove the maximum amount of the valuable mineral at as coarse a particle size as practical. The finer an ore is ground, the greater the energy that is required, so it makes sense to fine grind only those particles that need fine grinding. Complete liberation is not required for rougher flotation, only sufficient liberation to release enough gangue from the valuable mineral to get a high recovery.

The primary objective of roughing is to recover as much of the valuable minerals as possible, with less emphasis on the quality of the concentrate produced.

In some concentrators, there may be a *preflotation* step that precedes roughing. This is done when there are some undesirable materials, such as organic carbon, that readily float. They are removed first to avoid them floating during roughing (and thus contaminating the rougher concentrate).

Cleaning

The rougher concentrate is normally subjected to further stages of flotation to reject more of the undesirable minerals that also reported to the froth, in a process known as *cleaning*. The product of cleaning is known as the *cleaner concentrate* or the *final concentrate*.

The objective of cleaning is to produce as high a concentrate grade as possible.

The rougher concentrate is often subject to further grinding (usually called *regrinding*) to get more complete liberation of the valuable minerals. Because it is a smaller mass than that of the original ore, less energy is needed than would be necessary if the whole ore were reground. Regrinding is often undertaken in specialized *regrind mills*, such as the IsaMill™, designed to further reduce the energy consumed during regrinding to finer sizes.

Scavenging

The rougher flotation step is often followed by a *scavenger* flotation step that is applied to the rougher tailings. The objective is to recover any of the target minerals that were not recovered during the initial roughing stage. This might be achieved by changing the flotation conditions to make them more rigorous than the initial roughing, or there might be some secondary grinding to provide further liberation.

The concentrate from the rougher scavengers could be returned to the rougher feed for refloating or sent to special cleaner cells.

Similarly, the cleaning step may also be followed by a scavenging step performed on the cleaner tailings.

Science of Flotation

To be effective on a given ore slurry, the collectors are chosen based upon their selective wetting of the types of particles to be separated. A good collector will adsorb, physically or chemically, with one of the types of particles. This provides the thermodynamic requirement for the particles to bind to the surface of a bubble. The wetting activity of a surfactant on a particle can be quantified by measuring the contact angles that the liquid/bubble interface makes with it. Another important measure for attachment of bubbles to particles is induction time. The induction time is the time required for the particle and bubble to rupture the thin film separating the particle and bubble. This rupturing is achieved by the surface forces between the particle and bubble.

The mechanisms for the bubble-particle attachment is very complex and consists of three steps, collision, attachment and detachment. The collision is achieved by particles being within the collision tube of a bubble and this is affected by the velocity of the bubble and radius of the bubble. The collision tube corresponds to the region in which a particle will collide with the bubble, with the perimeter of the collision tube corresponding to the grazing trajectory.

The attachment of the particle to the bubble is controlled by the induction time of the particle and bubble. The particle and bubble need to bind and this occurs if the time in which the particle and bubble are in contact with each other is larger than the required induction time. This induction time is affected by the fluid viscosity, particle and bubble size and the forces between the particle and bubbles.

The detachment of a particle and bubble occurs when the force exerted by the surface tension is exceeded by shear forces and gravitational forces. These forces are complex and vary within the cell. High shear will be experienced close to the impeller of a mechanical flotation cell and mostly gravitational force in the collection and cleaning zone of a flotation column.

Significant issues of entrainment of fine particles occurs as these particles experience low collision efficiencies as well as sliming and degradation of the particle surfaces. Coarse particles show a low recovery of the valuable mineral due to the low liberation and high detachment efficiencies.

Theory

Selective Adhesion

Froth flotation depends on the selective adhesion of air bubbles to mineral surfaces in a mineral/water slurry. The air bubbles will attach to more hydrophobic particles. The attachment of the bubbles to the surface is determined by the interfacial energies between the solid, liquid, and gas phases. This is determined by the Young-Dupré Equation:

$$\gamma_{lv}\cos\theta = (\gamma_{sv} - \gamma_{sl})$$

where:

- γ_{lv} is the surface energy of the liquid/vapor interface

- γ_{sv} is the surface energy of the solid/vapor interface

- γ_{sl} is the surface energy of the solid/liquid interface,

- θ is the contact angle, the angle formed at the junction between vapor, solid, and liquid phases.

Minerals targeted for separation may be chemically surface-modified with collectors so that they are more hydrophobic. Collectors are a type of surfactant that increase the natural hydrophobicity of the surface, increasing the separability of the hydrophobic and hydrophilic particles. Collectors either chemically bond via chemisorption to the mineral or adsorb onto the surface via physisorption.

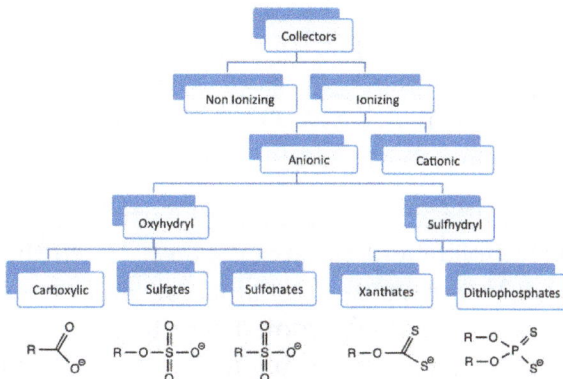

Different types of collectors, or surfactants, used in froth flotation.

IMFs and surface forces in Bubble-particle Interactions

Collision

The collision rates for fine particles (50 - 80 μm) can be accurately modeled, but there is no current theory that accurately models bubble-particle collision for particles as large as 300 μm, which are commonly used in flotation processes.

For fine particles, Stokes law underestimates collision probability while the potential equation based on Surface Charge overestimates collision probability so an intermediate equation is used.

It is important to know the collision rates in the system since this step precedes the adsorption where a three phase system is formed.

Adsorption (Attachment)

The effectiveness of a medium to adsorb to a particle is influenced by the relationship between the surfaces both materials. There are multiple factors that affect the efficiency of adsorption in chemical, thermodynamic, and physical domains. These factors can range from surface energy and polarity to the shape, size, and roughness of the particle. In froth flotation, adsorption is a strong consequence of surface energy, since the small particles have a high surface area to size ratio, resulting in higher energy surfaces to form attractions with adsorbates. The air bubbles must selectively adhere to the desired minerals to elevate them to the surface of the slurry while wetting the other minerals and leaving them in the aqueous slurry medium.

Particles that can be easily wetted by water are called hydrophilic, while particles that are not easily wetted by water are called hydrophobic. Hydrophobic particles have a tendency to form a separate phase in aqueous media. In froth flotation the effectiveness of an air bubble to adhere to a particle is based on how hydrophobic the particle is. Hydrophobic particles have an affinity to air bubbles, leading to adsorption. The bubble-particle combinations are elevated to the froth zone driven by buoyancy forces.

The attachment of the bubbles to the particles is determined by the interfacial energies of between the solid, liquid, and vapor phases, as modeled by the Young/Dupre Equation. The interfacial energies can be based on the natural structure of the materials, or the addition of chemical treatments can improve energy compatibility.

Collectors are the main additives used to improve particle surfaces. They function as surfactants to selectively isolate and aid adsorption between the particles of interest and bubbles rising through the slurry. Common collectors used in flotation are anionic sulfur ligands, which have a bifunctional structure with an ionic portion which shares attraction with metals, and a hydrophobic portion such as a long hydrocarbon tail. These collectors coat a particle's surface with a monolayer of non-polar substance to aid separation from the aqueous phase by decreasing the adsorbed particle solubility in water. The adsorbed ligands can form micelles around the particles and form small-particle colloids improving stability and phase separation further.

Desorption (Detachment)

The adsorption of particles to bubbles is essential to separating the minerals from the slurry, but the minerals must be purified from the additives used in separation, such as the collectors, frothers, and modifiers. The product of the cleaning, or desorption process, is known as the cleaner concentrate. The detachment of a particle and bubble requires adsorption bond cleavage driven by shear forces. Depending on the flotation cell type, shear forces are applied by a variety of mechanical systems. Among the most common are impellers and mixers. Some systems combine the functionalities of these components by placing them at key locations where they can take part in multiple froth flotation mechanisms. Cleaning cells also take advantage of gravitational forces to improve separation efficiency.

Performance Calculations

Relevant Equations

A common quantity used to describe the collection efficiency of a froth flotation process is *flotation recovery* (R). This quantity incorporates the probabilities of collision and attachment of particles to gas flotation bubbles.

$$R = \frac{N_c}{\left(\frac{\pi}{4}\right)\left(d_p + d_b\right)^2 Hc}$$

where:

- $N_c = PN_c^i$, which is the product of the probability of the particle being collected (P) and the number of possible particle collisions (N_c^i)

- d_p is particle diameter

- d_b is bubble diameter

- H is a specified height within the flotation which the recovery was calculated

- c is the particle concentration

The following, are several additional mathematical methods often used to evaluate the effectiveness of froth flotation processes. These equations are more simple than the calculation for *flotation recovery*, as they are based solely on the amounts of inputs and outputs of the processes.

For the following equations:

- F is the weight percent of feed

- C is the weight percent concentrate

- T is the weight percent of tailings

- c, t, and f are the Metallurgical assays of the concentrate, tailings, and feed, respectively

Ratio of feed weight to concentrate weight $\frac{F}{C}$ (unitless)

$$\frac{F}{C} = \frac{c-t}{f-t}$$

Percent of metal recovered (X_R) in wt%

$$X_R = 100 \left(\frac{c}{f}\right)\left(\frac{f-t}{c-t}\right)$$

Percent of metal lost (X_L) in wt%

$$X_L = 100 - X_R$$

Percent of weight recovered $\left(X_W\right)$ in wt%

$$X_W = 100\left(\frac{C}{F}\right) = 100\frac{f-t}{c-t}$$

Grade-recovery Curves

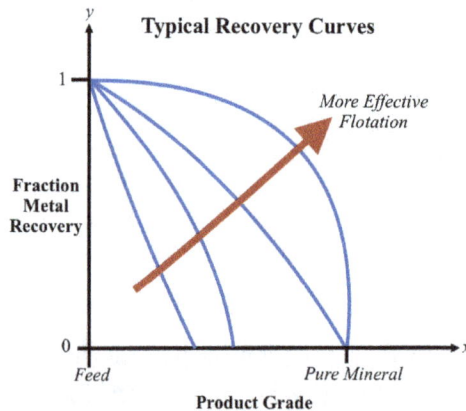

Example grade-recovery relationships seen in froth flotation.
Shifts in the curves represent changes in flotation effectiveness.

Grade-recovery curves are useful tools in weighing the trade-off of producing a high grade of concentrate while maintaining as low of a recovery rate as possible, two important aspects of froth flotation. These curves are developed empirically based on the individual froth flotation process of a particular plant. As the curves are shifted in the positive x-direction (to the right) and the positive y-direction (upward) the performance of the froth flotation process is regarded as improving. A disadvantage to these curves is that they can only compare the grade-recovery relations of a specific feed grade and feed rate. If a company has a variance of feed grades and rates used (an extremely common occurrence) in their froth flotation process, grade-recovery curves for every pairing of feed grade and recovery rate would have to be constructed in order to provide meaningful information to the plant.

Flotation Equipment

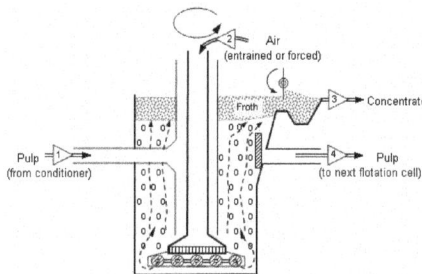

Diagram of froth flotation cell. Numbered triangles show direction of stream flow. A mixture of ore and water called pulp enters the cell from a conditioner, and flows to the bottom of the cell. Air or nitrogen is passed down a vertical impeller where shearing forces break the air stream into small bubbles. The mineral concentrate froth is collected from the top of the cell, while the pulp flows to another cell.

Flotation can be performed in rectangular or cylindrical mechanically agitated cells or tanks, flotation columns, Jameson Cells or deinking flotation machines. Classified by the method of air absorption manner, it is fair to state that two distinct groups of flotation equipments have arisen:pneumatic and mechanical machines. Generally pneumatic machines give a low-grade concentrate and little operating troubles.

Comparison of the sizes of flotation columns and Jameson Cells with similar capacities.

Mechanical cells use a large mixer and diffuser mechanism at the bottom of the mixing tank to introduce air and provide mixing action. Flotation columns use air spargers to introduce air at the bottom of a tall column while introducing slurry above. The countercurrent motion of the slurry flowing down and the air flowing up provides mixing action. Mechanical cells generally have a higher throughput rate, but produce material that is of lower quality, while flotation columns generally have a low throughput rate but produce higher quality material.

The Jameson cell uses neither impellers nor spargers, instead combining the slurry with air in a downcomer where high shear creates the turbulent conditions required for bubble particle contacting.

Mechanics of Flotation

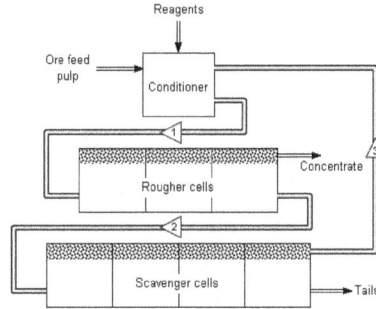

The following steps are followed, following grinding to liberate the mineral particles:

1. Reagent conditioning to achieve hydrophobic surface charges on the desired particles

2. Collection and upward transport by bubbles in an intimate contact with air or nitrogen

3. Formation of a stable froth on the surface of the flotation cell

4. Separation of the mineral laden froth from the bath (flotation cell)

Simple flotation circuit for mineral concentration. Numbered triangles show direction of stream flow. Various flotation reagents are added to a mixture of ore and water (called pulp) in a conditioning tank. The flow rate and tank size are designed to give the minerals enough time to be activated. The conditioner pulp is fed to a bank of rougher cells which remove most of the desired minerals as a concentrate. The rougher pulp passes to a bank of scavenger cells where additional reagents may be added. The scavenger cell froth is usually returned to the rougher cells for additional treatment, but in some cases may be sent to special cleaner cells. The scavenger pulp is usually barren enough to be discarded as tails. More complex flotation circuits have several sets of cleaner and re-cleaner cells, and intermediate re-grinding of pulp or concentrate.

Chemicals of Flotation

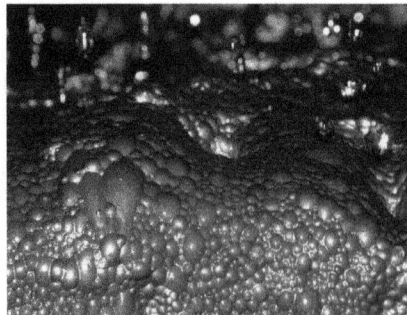

Copper sulfide foam in a froth-flotation cell

Collectors

For many ores (e.g. those of Cu, Mo, W, Ni), the collectors are anionic sulfur ligands. Particularly popular are xanthate salts, including potassium amyl xanthate (PAX), potassium isobutyl xanthate (PIBX), potassium ethyl xanthate (KEX), sodium isobutyl xanthate (SIBX), sodium isopropyl xanthate (SIPX), sodium ethyl xanthate (SEX). Other collectors include related sulfur-based ligands: dithiophosphates, dithiocarbamates. Still other classes of collectors include the thioureathiocarbanilide. Fatty acids have also been used.

For some minerals (e.g., sylvanite for KCl), fatty amines are used as collectors.

Frothers

A variety of compounds are added to stabilize the foams. These additives include pine oil, various alcohols (methyl isobutyl carbinol (MIBC)), polyglycols, xylenol (cresylic acid).

Modifiers

A variety of other compounds are added to optimize the separation process, these additives are called modifiers. Such species do not interact directly with the mineral, but modify the physical properties of the solution.

- pH modifiers include lime (CaO), Soda ash (Na_2CO_3), Caustic soda (NaOH), sulfuric and hydrochloric acid (H_2SO_4, HCl).

- Anionic modifiers include phosphates, silicates, and carbonates.

- Organic modifiers include the thickeners dextrin, starch, glue, and CMC.

Chemical Compounds for Deinking of Recycled Paper

- pH control: sodium silicate and sodium hydroxide

- Calciumion source: hard water, lime or calcium chloride

- Collector: fatty acid, fatty acid emulsion, fatty acid soap and/or organo-modified siloxane

Specific ore Applications

Illustrative, the flotation process is used for purification of potassium chloride from sodium chloride and clay minerals. The crushed mineral is suspended in brine in the presence of fatty ammonium salts. Because the ammonium group and K^+ have very similar ionic radii (ca. 0.135, 0.143 nm respectively), the ammonium centers exchange for the surface potassium sites on the particles of KCl, but not on the NaCl particles.

The long alkyl chains then confer hydrophobicity, to the particles, which enable them to form foams.

Sulfide ores		
• Copper	• Copper-Molybdenum	• Lead-Zinc
• Lead-Zinc-Iron	• Copper-Lead-Zinc-Iron	• Gold-Silver
• Oxide Copper and Lead	• Nickel	• Nickel-Copper
Nonsulfide ores		
• Fluorite	• Tungsten	• Lithium
• Tantalum	• Tin	• Coal

References

- Nguyen, Anh V (12 June 1996). "On modelling of bubble–particle attachment probability in flotation". International Journal of Mineral Processing. 53: 225–249. doi:10.1016/S0301-7516(97)00073-2

- Kittel, C. Kroemer, H. (1980). Thermal Physics, (first edition by Kittel alone 1969), second edition, W. H. Freeman, San Francisco, ISBN 0-7167-1088-9, pp. 49, 227

- "CODATA Value: molar gas constant". The NIST Reference on Constants, Units, and Uncertainty. US National Institute of Standards and Technology. June 2015. Retrieved 2015-09-25. 2014 CODATA recommended values

- Tro, N. J. (2008). Chemistry. A Molecular Approach, Pearson/Prentice Hall, Upper Saddle River NJ, ISBN 0-13-100065-9, p. 246

- What's in a Name? Amount of Substance, Chemical Amount, and Stoichiometric Amount Carmen J. Giunta Journal of Chemical Education 2016 93 (4), 583-586 doi:10.1021/acs.jchemed.5b00690

- Atkins, P., de Paula, J. (1978/2010). Physical Chemistry, (first edition 1978), ninth edition 2010, Oxford University Press, Oxford UK, ISBN 978-0-19-954337-3, p. 54

- Osborne, Graeme (1981). "Guillaume Daniel Delprat". Australian Dictionary of Biography. Canberra: Australian National University. Retrieved 7 June 2012

Processes and Methods of Metallurgy

Mineral processing separates the ore and the valuable minerals. The section elaborates on mineral processing methods such as roasting and calcination. In roasting, the concentrated ore is heated at a very high temperature. However, in calcination, the ore is subjected to high temperature in absence of air. All the diverse principles of mineral processing have been carefully analyzed in this chapter.

Mineral Processing

An important aspect of any mineral processing study is an analysis of how material is distributed whenever streams split and combine. This knowledge is necessary when a flow sheet is being designed and is also essential when making studies of operating plants.

Materials Balance in Mineral Processing

It is based on the principle of conservation of matter. In general

Input – output=accumulation (1)

In a continuous system at steady state, there is no accumulation and hence

Input=output

In mineral processing operations, single input of feed (ore) produces a concentrate containing most of the valuable and the tailing containing gangue minerals. Thus

Tons of feed (M_F) = Tons of concentrate (M_c) + and tons of tailing (M_T) (2)

$$M_F = M_c + M_T$$ (3)

Let f is fraction of metal in feed, c and t are fraction of metal in concentrate and tailing respectively, then

$$fM_F = cM_c - t\,M_t$$ (4)

By 3 and 4 we can obtain

$$\frac{\text{Mass of feed}}{\text{Mass of concentrate}} = \frac{M_F}{M_C} = \frac{c-t}{f-t} \tag{5}$$

Plant recovery (R) is $\dfrac{M_C \times c}{M_F \times f} \times 100$ $\qquad\qquad$ (6)

By 5 and 6 we get

$$R = \frac{c\,(f-t)}{f\,(c-t)} \times 100 \tag{7}$$

In measurements of quantities we derived the relationship between percent solids (%x) and pulp densities, when water is used as a medium to make the pulp, (pulp and slurry are synonyms).

$$\%x = \frac{100_{\rho s}\,(\rho m - 1000)}{\rho m (\rho s - 1000)} \tag{8}$$

ρ_s = density of solid and ρ_m = density of slurry

Mass flow rate of dry solids in pulp (slurry)

$$M = \frac{F\rho_s\,(\rho m - 1000)}{(\rho_s - 1000)} \tag{9}$$

F = volumetric flow rate (m^3 / hr) and M = mass flow rate in kg / hr

$$M = \frac{F\rho m \%x}{100}\,kg\,/\,hr \tag{10}$$

Water Balance (Dilution Ratio)

Water is used in mineral processing

1. To transport solids in the circuit

2. To act as a medium for separation

Ball mills use ~ 35% water for milling and in the discharge water is further added for separation in solids by weight.

Most flotation operations are performed in between 25 – 40% solids by weight.

Some gravity concentration devices operate most efficiently on slurry containing 55 – 70% solids.

Roughly $20\text{m}^3/\text{min}$ of water is required for a plant treating 10000 tons of ore.

Two product formula is of great use in assessing water balances. In two product formula; feed is divided in two products, namely concentrate and tailing.

Consider a hydrocyclone fed with a slurry containing $f_s\%$ solids by weight and producing two products:-

Under flow containing u% solids by wt. and an overflow containing V% solids by weight.

Consider weight of solids/unit of time in feed, underflow and overflow arc M_F^1, M_U and Mv respectively, at equilibrium conditions of operation

$$M_F^1 = M_U + M_v \tag{11}$$

Dilution ratio is defined as $= \dfrac{100 - \%\text{solids}}{\%\text{solids}}$

Dilution ratio is of feed $= \dfrac{100 - f_s}{f_s} = f_s^1$

Dilution ratio of underflow $= \dfrac{100 - u}{u} = u^1 \tag{12}$

Dilution ratio of overflow $= \dfrac{100 - v}{v} = v^1$

Water balance on the cyclone: weight of water entering the cyclone must equal the weight leaving in two products output

$$M_F^1 \times f_s^1 = M_U u^1 + M_v v^1 \tag{13}$$

By 11 and 13 we get

$$\frac{M_U}{M_F} = \frac{(f_s^1 - v^1)}{u^1 - v^1} \tag{14}$$

If % solids are unknown, two product balance can be performed by using pulp densities slurry densities.

A balance of slurry weights

$$\frac{MF}{\%f_s} = \frac{M_U}{\%u} + \frac{Mv}{\%v} \tag{15}$$

By equation 8 in that $\%x = \%$ solids and 15 we get after simplification

$$\frac{M_F \rho m}{\rho m - 1000} = \frac{M_U \rho u}{(\rho u - 1000)} = \frac{M_v \rho_v}{(\rho v - 1000)} \tag{16}$$

On simplifying further we get.

$$\frac{M_U}{M_F} = \frac{(\rho_v - \rho_f)(\rho_u - 1000)}{(\rho_v - \rho_u)(\rho_f - 1000)} = \frac{(\rho_f - \rho_v)(\rho_u - 1000)}{(\rho_u - \rho_v)(\rho_f - 1000)} \qquad (17)$$

In equation 17, ρ_u is density of slurry of underflow and ρ_v is density of slurry of overflow.

Example

Consider separation of feed into underflow and overflow by a hydro cyclone. Feed is 1000tons / hr and the underflow is 70% of the feed. Determine the circulating load ratio.

In the hydro cyclone underflow is re circulated

$$\text{Re circulating load ratio} = \frac{\text{Mass recycled}}{\text{fresh feed}}$$

Material balance gives mass of underflow =700 tons and that of overflow is 300 tons. Every time 300 tons is the fresh feed.

$$\text{Re circulating load ratio} \frac{700}{300} = 2.33$$

If the feed stream slurry contains 35% solids by volume and 40% of the water is re-cycled, calculate concentration of solids in hydro cyclone products. Density of solid $= 3.215\text{tons} / \text{m}^3$.

Hydro cyclone products: underflow and overflow

Mass balance gives 700 tons underflow and 300 tons overflow

Volume of solid in feed = volume of solid in underflow+ volume of solid in overflow

$$\frac{1000}{3.215} = \frac{700}{3.215} + \frac{300}{3.215}$$

$$311\text{m}^3 = 217.7\text{m}^3 + 93.9\text{m}^3$$

$$\text{Volume of water in feed} = 311 \times \frac{65}{35} = 578\text{m}^3$$

$$\text{Volume of water in underflow} = 0.4 \times 578 = 231\text{m}^3$$

$$\text{Solids concentration in underflow} = \frac{217.7 \times 100}{217.7 + 231} = 48.5\%$$

$$\text{Volume of water in overflow} = 578 - 231 = 347\text{m}^3.$$

$$\text{Solids concentration in overflow} = \frac{93.3}{93.3 + 347} \times 100 = 21.2\%$$

Calcination

Authorities differ on the meaning of calcination (also referred to as calcining). The IUPAC defines it as 'heating to high temperatures in air or oxygen'. However, calcination is also used to mean a thermal treatment process in the absence or limited supply of air or oxygen applied to ores and other solid materials to bring about a thermal decomposition. A calciner is a steel cylinder that rotates inside a heated furnace and performs indirect high-temperature processing (550–1150 °C, or 1000–2100 °F) within a controlled atmosphere.

Industrial Processes

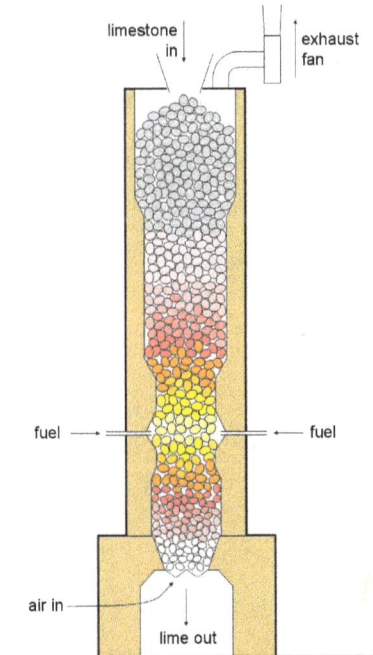

An oven for calcination of limestone

The process of calcination derives its name from the Latin *calcinare* (to burn lime) due to its most common application, the decomposition of calcium carbonate (limestone) to calcium oxide (lime) and carbon dioxide, in order to create cement. The product of calcination is usually referred to in general as "calcine", regardless of the actual minerals undergoing thermal treatment. Calcination is carried out in furnaces or reactors (sometimes referred to as kilns or calciners) of various designs including shaft furnaces, rotary kilns, multiple hearth furnaces, and fluidized bed reactors.

Examples of calcination processes include the following:

- decomposition of carbonate minerals, as in the calcination of limestone to drive off carbon dioxide;

- decomposition of hydrated minerals, as in the calcination of bauxite and gypsum, to remove crystalline water as water vapor;

- decomposition of volatile matter contained in raw petroleum coke;

- heat treatment to effect phase transformations, as in conversion of anatase to rutile or devitrification of glass materials

- removal of ammoniumions in the synthesis of zeolites.

Calcination Reactions

Calcination reactions usually take place at or above the thermal decomposition temperature (for decomposition and volatilization reactions) or the transition temperature (for phase transitions). This temperature is usually defined as the temperature at which the standard Gibbs free energy for a particular calcination reaction is equal to zero. For example, in limestone calcination, a decomposition process, the chemical reaction is

$$CaCO_3 \rightarrow CaO + CO_2(g)$$

The standard Gibbs free energy of reaction is approximated as $\Delta G^\circ_r = 177{,}100 - 158\,T$ (J/mol). The standard free energy of reaction is 0 in this case when the temperature, T, is equal to 1121 K, or 848 °C.

Oxidation

In some cases, calcination of a metal results in oxidation of the metal. Jean Rey noted that lead and tin when calcinated gained weight, presumably as they were being oxidized.

Alchemy

In alchemy, calcination was believed to be one of the 12 vital processes required for the transformation of a substance.

Alchemists distinguished two kinds of calcination, *actual* and *potential*. Actual calcination is that brought about by actual fire, from wood, coals, or other fuel, raised to a certain temperature. Potential calcination is that brought about by *potential* fire, such as corrosive chemicals; for example, gold was calcined in a reverberatory furnace with mercury and sal ammoniac; silver with common salt and alkali salt; copper with salt and sulfur; iron with sal ammoniac and vinegar; tin with antimony; lead with sulfur; and mercury with aqua fortis.

There was also *philosophical calcination*, which was said to occur when horns, hooves, etc., were hung over boiling water, or other liquor, until they had lost their mucilage, and were easily reducible into powder.

Roasting (Metallurgy)

Roasted gold ore from Cripple Creek, Colorado. Roasting has driven off the tellurium from the original calaverite, leaving behind vesicular blebs of native gold.

Roasting is a process of heating of concentrated ore to a high temperature in presence of air. It is a step of the processing of certain ores. More specifically, roasting is a metallurgical process involving gas–solid reactions at elevated temperatures with the goal of purifying the metal component(s). Often before roasting, the ore has already been partially purified, e.g. by froth flotation. The concentrate is mixed with other materials to facilitate the process. The technology is useful but is also a serious source of air pollution.

Roasting consists of thermal gas–liquid reactions, which can include oxidation, reduction, chlorination, sulfation, and pyrohydrolysis. In roasting, the ore or ore concentrate is treated with very hot air. This process is generally applied to sulphide minerals. During roasting, the sulphide is converted to an oxide, and sulphur is released as sulphur dioxide, a gas. For the ores Cu_2S (chalcocite) and ZnS (sphalerite), balanced equations for the roasting are:

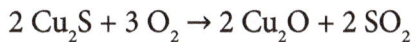

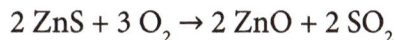

$$2\,Cu_2S + 3\,O_2 \rightarrow 2\,Cu_2O + 2\,SO_2$$

$$2\,ZnS + 3\,O_2 \rightarrow 2\,ZnO + 2\,SO_2$$

The gaseous product of sulfide roasting, sulfur dioxide (SO_2) is often used to produce sulfuric acid. Many sulfide minerals contain other components such as arsenic that are released into the environment.

Up until the early 20th century, roasting was started by burning wood on top of ore. This would raise the temperature of the ore to the point where its sulfur content would become its source of fuel, and the roasting process could continue without external fuel sources. Early sulfide roasting was practiced in this manner in "open hearth" roasters, which were manually stirred (a practice called "rabbling") using rake-like tools to expose unroasted ore to oxygen as the reaction proceeded.

This process released large amounts of acidic, metallic, and other toxic compounds. Results of this include areas that even after 60–80 years are still largely lifeless, often exactly corresponding to the area of the roast bed, some of which are hundreds of metres wide by kilometres long. Roasting is an exothermic process.

Roasting Operations

The following describe different forms of roasting

Oxidizing Roasting

Oxidizing roasting, the most commonly practiced roasting process, involves heating the ore in excess of air or oxygen, to burn out or replace the impurity element, generally sulfur, partly or completely by oxygen. For sulfide roasting, the general reaction can be given by -

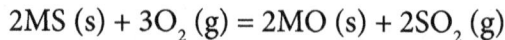

$$2MS\ (s) + 3O_2\ (g) = 2MO\ (s) + 2SO_2\ (g)$$

Roasting the sulfide ore, till almost complete removal of the sulfur from the ore, results in a *dead roast*.

Volatilizing Roasting

Volatilizing roasting, involves careful oxidation at elevated temperatures of the ores, to eliminate impurity elements in the form of their volatile oxides. Examples of such volatile oxides include As_2O_3, Sb_2O_3, ZnO and sulfur oxides. Careful control of the oxygen content in the roaster is necessary, as excessive oxidation forms non volatile oxides.

Chloridizing Roasting

Chloridizing roasting transforms certain metal compounds to chlorides, through oxidation or reduction. Some metals like uranium, titanium, beryllium and some rare earths are processed in their chloride form. Certain forms of chloridizing roasting maybe represented by the overall reactions -

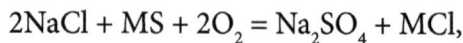

$$2NaCl + MS + 2O_2 = Na_2SO_4 + MCl,$$

$$4NaCl + 2MO + S_2 + 3O_2 = 2Na_2SO_4 + 2MCl_2$$

The first reaction represents the chlorination of a sulfide ore involving an exothermic reaction. The second reaction involving an oxide ore is facilitated by addition of elemental sulfur. Carbonate ores react in a similar manner as the oxide ore, after decomposing to their oxide form at high temperature.

Sulfating Roasting

Sulfating roasting oxidizes certain sulfide ores to sulfates in a controlled supply of air to enable leaching of the sulfate for further processing.

Magnetic Roasting

Magnetic roasting involves controlled roasting of the ore to convert it into a magnetic form, thus enabling easy separation and processing in subsequent steps. For example, controlled reduction of haematite (non magnetic Fe_2O_3) to magnetite (magnetic Fe_3O_4).

Reduction Roasting

Reduction roasting partially reduces an oxide ore before the actual smelting process.

Sinter Roasting

Sinter roasting involves heating the fine ores at high temperatures, where simultaneous oxidation and agglomeration of the ores take place. For example, lead sulfide ores are subjected to sinter roasting in a continuous process after froth flotation to convert the fine ores to workable agglomerates for further smelting operations.

Sources of Energy

Different sources of thermal energy are

1. Heat of reaction: Combustion of S to SO_2 or SO_3 releases -70940 k cal / kgmol and -93900 k cal / kg mol of thermal energy respectively. Several oxidation reactions like $Fe-Fe_2O_3, FeS \rightarrow FeO, ZnS-ZnO$ and so on releases thermal energy. These energies can be calculated from the heat of formation. Heat of formation of some compounds is

Compound	$\Delta H_f^0, 298$ (kcal/kg mol)
Cu_2S	-18950
ZnS	-44000
FeS_2	-35500
FeS	-23100
CO_2	-94450
CO	-26840
Cu_2O	-42500
CuO	-38500
ZnO	- 83500
PbO	-52500

For Example

$$Cu_2S(c) + 2O_2(g) = 2\ CuO(c) + SO_2(g)$$

(c) is condensed and (g) is gaseous phase

This reaction produces heat at 298K

$$-\Delta H_R^o = 2 \times (\Delta H_f^o)_{CuO} + (\Delta H_f^o)_{SO_2} - (\Delta H_f^o)_{Cu_2S}$$

$$-\Delta H_R^0 = 136990\ k\,cal$$

$$ZnS + 1.5\ O_2 = ZnO + SO_2$$

$$\Delta H_R^o = (\Delta H_f^o)_{ZnO} + (\Delta H_f^o)_{SO_2} - \Delta H_{f(ZnS)}^o$$

$$\Delta H_R^o = -110440\ k\,cal$$

$$PbS + 2O_2 = PbSO_4$$

$$\Delta H_R^o = -197000\ kcal$$

Combustion of Fuel

In the ore concentrate solid fuel is mixed for roasting. Solid fuel contains carbon and hydrogen besides other elements. Fuel is characterized by calorific value. Gross calorific value of a solid fuel can be calculated by

Gross calorific value $(Kcal / kg)(GCV)$

$$(GCV) = 81\%C + 341\left(\%H - F\frac{\%O}{8}\right) + 22\%S$$

If the coal analyses $C74\%, H6\%, N1\%, 09\%, S0.8\%$ moisture 2.2% and ash 8%, GCV of coal is $32060\ kJ / kg$.

Combustion

Combustion or burning is a high-temperature exothermicredoxchemical reaction between a fuel (the reductant) and an oxidant, usually atmospheric oxygen, that produces oxidized, often gaseous products, in a mixture termed as smoke. Combustion in a fire produces a flame, and the heat produced can make combustion self-sustaining. Combustion is often a complicated sequence of elementaryradical reactions. Solid fuels, such as wood, first undergo endothermicpyrolysis to produce gaseous fuels whose combustion then supplies the heat required to produce more of them. Combustion is often hot enough that light in the form of either glowing or a flame is produced. A simple example can be seen in the combustion of hydrogen and oxygen into water vapor, a reaction commonly used to fuel rocket engines. This reaction releases 242 kJ/mol of heat and reduces the enthalpy accordingly (at constant temperature and pressure):

The flames caused as a result of a fuel undergoing combustion (burning)

Air pollution abatement equipment provides combustion control for industrial processes.

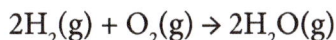

$$2H_2(g) + O_2(g) \rightarrow 2H_2O(g)$$

Combustion of an organic fuel in air is always exothermic because the double bond in O_2 is much weaker than other double bonds or pairs of single bonds, and therefore the formation of the stronger bonds in the combustion products CO_2 and H_2O results in the release of energy. The bond energies in the fuel play only a minor role, since they are similar to those in the combustion products; e.g., the sum of the bond energies of CH_4 is nearly the same as that of CO_2. The heat of combustion is approximately -418 kJ per mole of O_2 used up in the combustion reaction, and can be estimated from the elemental composition of the fuel.

Uncatalyzed combustion in air requires fairly high temperatures. Complete combustion is stoichiometric with respect to the fuel, where there is no remaining fuel, and ideally, no remaining oxidant. Thermodynamically, the chemical equilibrium of combustion in air is overwhelmingly on the side of the products. However, complete combustion is almost impossible to achieve, since the chemical equilibrium is not necessarily reached, or may contain unburnt products such as carbon monoxide, hydrogen and even carbon (soot or ash). Thus, the produced smoke is usually toxic and contains unburned or partially oxidized products. Any combustion at high temperatures in atmosphericair, which is 78 percent nitrogen, will also create small amounts of several nitrogen oxides, commonly referred to as NOx, since the combustion of nitrogen is thermodynamically favored at high, but not low temperatures. Since combustion is rarely clean, flue gas cleaning or catalytic converters may be required by law.

Fires occur naturally, ignited by lightning strikes or by volcanic products. Combustion (fire) was the first controlled chemical reaction discovered by humans, in the form of campfires and bonfires, and continues to be the main method to produce energy for humanity. Usually, the fuel is carbon, hydrocarbons or more complicated mixtures such as wood that contains partially oxidized hydrocarbons. The thermal energy produced from combustion of either fossil fuels such as coal or oil, or from renewable fuels such as firewood, is harvested for diverse uses such as cooking, production of electricity or industrial or domestic heating. Combustion is also currently the only reaction used to power rockets. Combustion is also used to destroy (incinerate) waste, both nonhazardous and hazardous.

Oxidants for combustion have high oxidation potential and include atmospheric or pure oxygen, chlorine, fluorine, chlorine trifluoride, nitrous oxide and nitric acid. For instance, hydrogen burns in chlorine to form hydrogen chloride with the liberation of heat and light characteristic of combustion. Although usually not catalyzed, combustion can be catalyzed by platinum or vanadium, as in the contact process.

Types

Complete

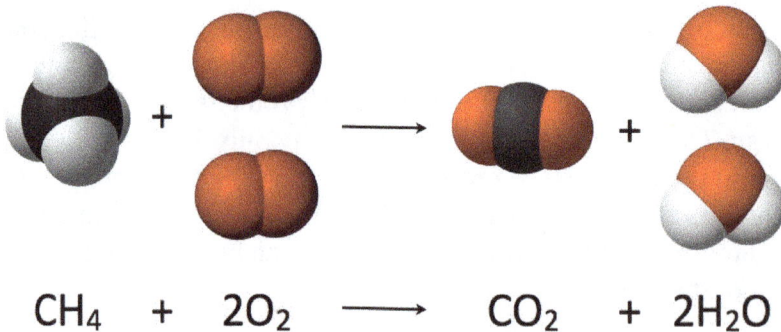

$$CH_4 + 2O_2 \longrightarrow CO_2 + 2H_2O$$

The combustion of methane, a hydrocarbon.

In complete combustion, the reactant burns in oxygen, producing a limited number of products. When a hydrocarbon burns in oxygen, the reaction will primarily yield carbon dioxide and water. When elements are burned, the products are primarily the most common oxides. Carbon will yield carbon dioxide, sulfur will yield sulfur dioxide, and iron will yield iron(III) oxide. Nitrogen is not considered to be a combustible substance when oxygen is the oxidant, but small amounts of various nitrogen oxides (commonly designated NOx species) form when air is the oxidant.

Combustion is not necessarily favorable to the maximum degree of oxidation, and it can be temperature-dependent. For example, sulfur trioxide is not produced quantitatively by the combustion of sulfur. NOx species appear in significant amounts above about 2,800 °F (1,540 °C), and more is produced at higher temperatures. The amount of NOx is also a function of oxygen excess.

In most industrial applications and in fires, air is the source of oxygen (O_2). In air, each mole of oxygen is mixed with approximately 3.71 mol of nitrogen. Nitrogen does not take part in combustion, but at high temperatures some nitrogen will be converted to NOx (mostly NO, with much smaller amounts of NO_2). On the other hand, when there is insufficient oxygen to completely combust the fuel, some fuel carbon is converted to carbon monoxide and some of the hydrogen remains unreacted. A more complete set of equations for the combustion of a hydrocarbon in air therefore requires an additional calculation for the distribution of oxygen between the carbon and hydrogen in the fuel.

The amount of air required for complete combustion to take place is known as theoretical air. However, in practice the air used is 2-3x that of theoretical air.

Incomplete

Incomplete combustion will occur when there is not enough oxygen to allow the fuel to react completely to produce carbon dioxide and water. It also happens when the combustion is quenched by a heat sink, such as a solid surface or flame trap.

For most fuels, such as diesel oil, coal or wood, pyrolysis occurs before combustion. In incomplete combustion, products of pyrolysis remain unburnt and contaminate the smoke with noxious particulate matter and gases. Partially oxidized compounds are also a concern; partial oxidation of ethanol can produce harmful acetaldehyde, and carbon can produce toxic carbon monoxide.

The quality of combustion can be improved by the designs of combustion devices, such as burners and internal combustion engines. Further improvements are achievable by catalytic after-burning devices (such as catalytic converters) or by the simple partial return of the exhaust gases into the combustion process. Such devices are required by environmental legislation for cars in most countries, and may be necessary to enable large combustion devices, such as thermal power stations, to reach legal emission standards.

The degree of combustion can be measured and analyzed with test equipment. HVAC contractors, firemen and engineers use combustion analyzers to test the efficiency of a burner during the combustion process. In addition, the efficiency of an internal combustion engine can be measured in this way, and some U.S. states and local municipalities use combustion analysis to define and rate the efficiency of vehicles on the road today.

Smouldering

Smouldering is the slow, low-temperature, flameless form of combustion, sustained by the heat evolved when oxygen directly attacks the surface of a condensed-phase fuel. It is a typically incomplete combustion reaction. Solid materials that can sustain a smouldering reaction include coal, cellulose, wood, cotton, tobacco, peat, duff, humus, synthetic foams, charring polymers (including polyurethane foam) and dust. Common examples of smouldering phenomena are the initiation of residential fires on upholstered furniture by weak heat sources (e.g., a cigarette, a short-circuited wire) and the persistent combustion of biomass behind the flaming fronts of wildfires.

Rapid

Rapid combustion is a form of combustion, otherwise known as a fire, in which large amounts of heat and light energy are released, which often results in a flame. This is used in a form of machinery such as internal combustion engines and in thermobaric weapons. Such a combustion is frequently called an explosion, though for an internal combustion engine this is inaccurate. An internal combustion engine nominally operates on a controlled rapid burn. When the fuel-air mixture in an internal combustion engine explodes, that is known as detonation.

Spontaneous

Spontaneous combustion is a type of combustion which occurs by self heating (increase in temperature due to exothermic internal reactions), followed by thermal runaway (self heating which rapidly accelerates to high temperatures) and finally, ignition. For example, phosphorus self-ignites at room temperature without the application of heat.

Turbulent

Combustion resulting in a turbulent flame is the most used for industrial application (e.g. gas turbines, gasoline engines, etc.) because the turbulence helps the mixing process between the fuel and oxidizer.

Micro-gravity

The term 'micro' gravity refers to a gravitational state that is 'low' (i.e., 'micro' in the sense of 'small' and not necessarily a millionth of Earth's normal gravity) such that the influence of buoyancy on physical processes may be considered small relative to other flow processes that would be present at normal gravity. In such an environment, the thermal and flow transport dynamics can behave quite differently than in normal gravity conditions (e.g., a candle's flame takes the shape of a sphere.). Microgravity combustion research contributes to the understanding of a wide variety of aspects that are relevant to both the environment of a spacecraft (e.g., fire dynamics relevant to crew safety on the International Space Station) and terrestrial (Earth-based) conditions (e.g., droplet combustion dynamics to assist developing new fuel blends for improved combustion, materials fabrication processes, thermal management of electronic systems, multiphase flow boiling dynamics, and many others).

Colourized gray-scale composite image of the individual frames from a video of a backlit fuel droplet burning in microgravity.

Micro-combustion

Combustion processes which happen in very small volumes are considered micro-combustion. The high surface-to-volume ratio increases specific heat loss. Quenching distance plays a vital role in stabilizing the flame in such combustion chambers.

Chemical Equations

Stoichiometric Combustion of a Hydrocarbon in Oxygen

Generally, the chemical equation for stoichiometric combustion of a hydrocarbon in oxygen is:

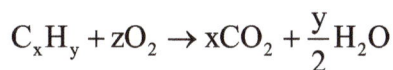

$$C_xH_y + zO_2 \rightarrow xCO_2 + \frac{y}{2}H_2O$$

where $z = x + y/4$.

For example, the stoichiometric burning of propane in oxygen is:

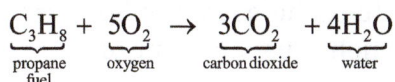

$$\underbrace{C_3H_8}_{\substack{\text{propane} \\ \text{fuel}}} + \underbrace{5O_2}_{\text{oxygen}} \rightarrow \underbrace{3CO_2}_{\text{carbon dioxide}} + \underbrace{4H_2O}_{\text{water}}$$

Stoichiometric Combustion of a Hydrocarbon in Air

If the stoichiometric combustion takes place using air as the oxygen source, the nitrogen present in the air (Atmosphere of Earth) can be added to the equation (although it does not react) to show the stoichiometric composition of the fuel in air and the composition of the resultant flue gas. Note that treating all non-oxygen components in air as nitrogen gives a 'nitrogen' to oxygen ration of 3.77, i.e. (100% - O2%) / O2% where O2% is 20.95% vol:

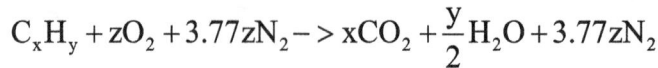

$$C_xH_y + zO_2 + 3.77zN_2 -> xCO_2 + \frac{y}{2}H_2O + 3.77zN_2$$

where $z = x + \frac{1}{4}y$.

For example, the stoichiometric combustion of propane (C3H8) in air is:

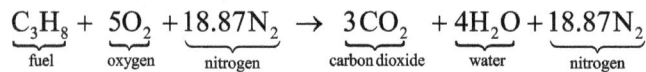

$$\underbrace{C_3H_8}_{\text{fuel}} + \underbrace{5O_2}_{\text{oxygen}} + \underbrace{18.87N_2}_{\text{nitrogen}} \rightarrow \underbrace{3CO_2}_{\text{carbon dioxide}} + \underbrace{4H_2O}_{\text{water}} + \underbrace{18.87N_2}_{\text{nitrogen}}$$

The stoichiometric composition of propane in air is 1 / (1 + 5 + 18.87) = 4.02% vol.

Trace Combustion Products

Various other substances begin to appear in significant amounts in combustion products when the flame temperature is above about 1600 K. When excess air is used, nitrogen may oxidize to NO and, to a much lesser extent, to NO_2. CO forms by disproportionation of CO_2, and H_2 and OH form by disproportionation of H_2O.

For example, when 1 mol of propane is burned with 28.6 mol of air (120% of the stoichiometric amount), the combustion products contain 3.3% O_2. At 1400 K, the equilibrium combustion products contain 0.03% NO and 0.002% OH. At 1800 K, the combustion products contain 0.17% NO, 0.05% OH, 0.01% CO, and 0.004% H_2.

Diesel engines are run with an excess of oxygen to combust small particles that tend to form with only a stoichiometric amount of oxygen, necessarily producing nitrogen oxide emissions. Both the United States and European Union enforce limits to vehicle nitrogen oxide emissions, which necessitate the use of special catalytic converters or treatment of the exhaust with urea.

Incomplete Combustion of a Hydrocarbon in Oxygen

The incomplete (partial) combustion of a hydrocarbon with oxygen produces a gas mixture containing mainly CO_2, CO, H_2O, and H_2. Such gas mixtures are commonly prepared for use as protective atmospheres for the heat-treatment of metals and for gas carburizing. The general reaction equation for incomplete combustion of one mole of a hydrocarbon in oxygen is:

$$\underbrace{C_xH_y}_{\text{fuel}} + \underbrace{ZO_2}_{\text{oxygen}} \rightarrow \underbrace{aCO_2}_{\text{carbon dioxide}} + \underbrace{bCO}_{\text{carbon monoxide}} + \underbrace{cH_2O}_{\text{water}} + \underbrace{dH_2}_{\text{hydrogen}}$$

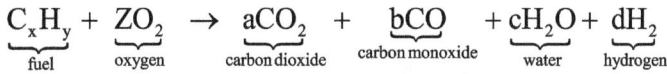

When z falls below roughly 50% of the stoichiometric value, CH_4 can become an important combustion product; when z falls below roughly 35% of the stoichiometric value, elemental carbon may become stable.

The products of incomplete combustion can be calculated with the aid of a material balance, together with the assumption that the combustion products reach equilibrium. For example, in the combustion of one mole of propane (C_3H_8) with four moles of O_2, seven moles of combustion gas are formed, and z is 80% of the stoichiometric value. The three elemental balance equations are:

- Carbon: $a + b = 3$

- Hydrogen: $2c + 2d = 8$

- Oxygen: $2a + b + c = 8$

These three equations are insufficient in themselves to calculate the combustion gas composition. However, at the equilibrium position, the water-gas shift reaction gives another equation:

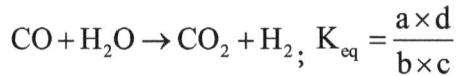

$$CO + H_2O \rightarrow CO_2 + H_2; \quad K_{eq} = \frac{a \times d}{b \times c}$$

For example, at 1200 K the value of K_{eq} is 0.728. Solving, the combustion gas consists of 42.4% H_2O, 29.0% CO_2, 14.7% H_2, and 13.9% CO. Carbon becomes a stable phase at 1200 K and 1 atm pressure when z is less than 30% of the stoichiometric value, at which point the combustion products contain more than 98% H_2 and CO and about 0.5% CH_4.

Fuels

Substances or materials which undergo combustion are called fuels. The most common examples are natural gas, propane, kerosene, diesel, petrol, charcoal, coal, wood, etc.

Liquid Fuels

Combustion of a liquid fuel in an oxidizing atmosphere actually happens in the gas phase. It is the vapor that burns, not the liquid. Therefore, a liquid will normally catch fire only above a certain temperature: its flash point. The flash point of a liquid fuel is the lowest temperature at which it can form an ignitable mix with air. It is the minimum temperature at which there is enough evaporated fuel in the air to start combustion.

Gaseous Fuels

Combustion of gaseous fuels may occur through one of four distinctive types of burning: diffusion flame, premixed flame, autoignitive reaction front, or as a detonation. The type of burning that actually occurs depends on the degree to which the fuel and oxidizer are mixed prior to heating: for example, a diffusion flame is formed if the fuel and oxidizer are separated initially, whereas a premixed flame is formed otherwise. Similarly, the type of burning also depends on the pressure: a detonation, for example, is an autoignitive reaction front coupled to a strong shock wave giving it its characteristic high-pressure peak and high detonation velocity.

Solid Fuels

The act of combustion consists of three relatively distinct but overlapping phases:

- Preheating phase, when the unburned fuel is heated up to its flash point and then fire point. Flammable gases start being evolved in a process similar to dry distillation.

- Distillation phase or gaseous phase, when the mix of evolved flammable gases with oxygen is ignited. Energy is produced in the form of heat and light. Flames are often visible. Heat transfer from the combustion to the solid maintains the evolution of flammable vapours.

- Charcoal phase or solid phase, when the output of flammable gases from the material is too low for persistent presence of flame and the charred fuel does not burn rapidly and just glows and later only smoulders.

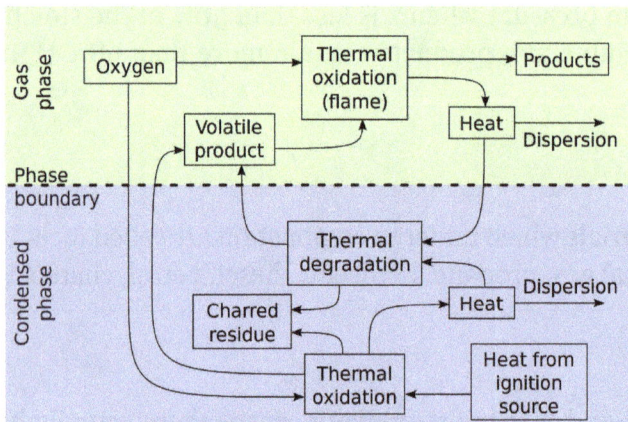

A general scheme of polymer combustion

Combustion Management

Efficient process heating requires recovery of the largest possible part of a fuel's heat of combustion into the material being processed. There are many avenues of loss in the operation of a heating process. Typically, the dominant loss is sensible heat leaving with the offgas (i.e., the flue gas). The temperature and quantity of offgas indicates its heat content (enthalpy), so keeping its quantity low minimizes heat loss.

In a perfect furnace, the combustion air flow would be matched to the fuel flow to give each fuel molecule the exact amount of oxygen needed to cause complete combustion. However, in the real world, combustion does not proceed in a perfect manner. Unburned fuel (usually CO and H_2) discharged from the system represents a heating value loss (as well as a safety hazard). Since combustibles are undesirable in the offgas, while the presence of unreacted oxygen there presents minimal safety and environmental concerns, the first principle of combustion management is to provide more oxygen than is theoretically needed to ensure that all the fuel burns. For methane (CH_4) combustion, for example, slightly more than two molecules of oxygen are required.

The second principle of combustion management, however, is to not use too much oxygen. The correct amount of oxygen requires three types of measurement: first, active control of air and fuel flow; second, offgas oxygen measurement; and third, measurement of offgas combustibles. For each heating process there exists an optimum condition of minimal offgas heat loss with acceptable levels of combustibles concentration. Minimizing excess oxygen pays an additional benefit: for a given offgas temperature, the NOx level is lowest when excess oxygen is kept lowest.

Adherence to these two principles is furthered by making material and heat balances on the combustion process. The material balance directly relates the air/fuel ratio to the percentage of O2 in the combustion gas. The heat balance relates the heat available for the charge to the overall net heat produced by fuel combustion. Additional material and heat balances can be made to quantify the thermal advantage from preheating the combustion air, or enriching it in oxygen.

Reaction Mechanism

Combustion in oxygen is a chain reaction in which many distinct radical intermediates participate. The high energy required for initiation is explained by the unusual structure of the dioxygen molecule. The lowest-energy configuration of the dioxygen molecule is a stable, relatively unreactive diradical in a triplet spin state. Bonding can be described with three bonding electron pairs and two antibonding electrons, whose spins are aligned, such that the molecule has nonzero total angular momentum. Most fuels, on the other hand, are in a singlet state, with paired spins and zero total angular momentum. Interaction between the two is quantum mechanically a "forbidden transition", i.e. possible with a very low probability. To initiate combustion, energy is required to force dioxygen into a spin-paired state, or singlet oxygen. This intermediate is extremely reactive. The energy is supplied as heat, and the reaction then produces additional heat, which allows it to continue.

Combustion of hydrocarbons is thought to be initiated by hydrogen atom abstraction (not proton abstraction) from the fuel to oxygen, to give a hydroperoxide radical (HOO). This reacts further to give hydroperoxides, which break up to give hydroxyl radicals. There are a great variety of these processes that produce fuel radicals and oxidizing radicals. Oxidizing species include singlet oxygen, hydroxyl, monatomic oxygen, and hydroperoxyl. Such intermediates are short-lived and cannot be isolated. However, non-radical intermediates are stable and are produced in incomplete combustion. An example is acetaldehyde produced in the combustion of ethanol. An intermediate in the combustion of carbon and hydrocarbons, carbon monoxide, is of special importance because it is a poisonous gas, but also economically useful for the production of syngas.

Solid and heavy liquid fuels also undergo a great number of pyrolysis reactions that give more easily oxidized, gaseous fuels. These reactions are endothermic and require constant energy input from the ongoing combustion reactions. A lack of oxygen or other poorly designed conditions result in these noxious and carcinogenic pyrolysis products being emitted as thick, black smoke.

The rate of combustion is the amount of a material that undergoes combustion over a period of time. It can be expressed in grams per second (g/s) or kilograms per second (kg/s).

Detailed descriptions of combustion processes, from the chemical kinetics perspective, requires the formulation of large and intricate webs of elementary reactions. For instance, combustion of hydrocarbon fuels typically involve hundreds of chemical species reacting according to thousands of reactions.

Inclusion of such mechanisms within computational flow solvers still represents a pretty challenging task mainly in two aspects. First, the number of degrees of freedom (proportional to the number of chemical species) can be dramatically large; second the source term due to reactions introduces a disparate number of time scales which makes the whole dynamical system stiff. As a result, the direct numerical simulation of turbulent reactive flows with heavy fuels soon becomes intractable even for modern supercomputers.

Therefore, a plethora of methodologies has been devised for reducing the complexity of combustion mechanisms without renouncing to high detail level. Examples are provided by: the Relaxation Redistribution Method (RRM) The Intrinsic Low-Dimensional Manifold (ILDM) approach and further developments The invariant constrained equilibrium edge preimage curve method. A few variational approaches The Computational Singular perturbation (CSP) method and further developments. The Rate Controlled Constrained Equilibrium (RCCE) and Quasi Equilibrium Manifold (QEM) approach. The G-Scheme. The Method of Invariant Grids (MIG).

Temperature

Antoine Lavoisier conducting an experiment related combustion generated by amplified sun light.

Assuming perfect combustion conditions, such as complete combustion under adiabatic conditions (i.e., no heat loss or gain), the adiabatic combustion temperature can be determined. The formula that yields this temperature is based on the first law of thermodynamics and takes note of the fact that the heat of combustion is used entirely for heating the fuel, the combustion air or oxygen, and the combustion product gases (commonly referred to as the *flue gas*).

In the case of fossil fuels burnt in air, the combustion temperature depends on all of the following:

- the heating value;
- the stoichiometric air to fuel ratio λ ;
- the specific heat capacity of fuel and air;
- the air and fuel inlet temperatures.

The adiabatic combustion temperature (also known as the *adiabatic flame temperature*) increases for higher heating values and inlet air and fuel temperatures and for stoichiometric air ratios approaching one.

Most commonly, the adiabatic combustion temperatures for coals are around 2,200 °C (3,992 °F) (for inlet air and fuel at ambient temperatures and for $\lambda = 1.0$), around 2,150 °C (3,902 °F) for oil and 2,000 °C (3,632 °F) for natural gas.

In industrial fired heaters, power stationsteam generators, and large gas-fired turbines, the more common way of expressing the usage of more than the stoichiometric combustion air is *percent excess combustion air*. For example, excess combustion air of 15 percent means that 15 percent more than the required stoichiometric air is being used.

Instabilities

Combustion instabilities are typically violent pressure oscillations in a combustion chamber. These pressure oscillations can be as high as 180 dB, and long term exposure to these cyclic pressure and thermal loads reduces the life of engine components. In rockets, such as the F1 used in the Saturn V program, instabilities led to massive damage of the combustion chamber and surrounding components. This problem was solved by re-designing the fuel injector. In liquid jet engines the droplet size and distribution can be used to attenuate the instabilities. Combustion instabilities are a major concern in ground-based gas turbine engines because of NOx emissions. The tendency is to run lean, an equivalence ratio less than 1, to reduce the combustion temperature and thus reduce the NOx emissions; however, running the combustion lean makes it very susceptible to combustion instability.

The Rayleigh Criterion is the basis for analysis of thermoacoustic combustion instability and is evaluated using the Rayleigh Index over one cycle of instability

$$G(x) = \frac{1}{T}\int_T q'(x,t)p'(x,t)dt$$

where q' is the heat release rate perturbation and p' is the pressure fluctuation. When the heat release oscillations are in phase with the pressure oscillations, the Rayleigh Index is positive and the magnitude of the thermo acoustic instability is maximised. On the other hand, if the Rayleigh Index is negative, then thermoacoustic damping occurs. The Rayleigh Criterion implies that a thermoacoustic instability can be optimally controlled by having heat release oscillations 180 degrees out of phase with pressure oscillations at the same frequency. This minimizes the Rayleigh Index.

Determination of Calorific Value of Gaseous Fuel

In determination of gaseous fuel, combustible components are: CO, H_2, hydrocarbons, NH_3 etc., whereas O_2, CO_2, N_2 are diluents.

Heats of Formation of Some Oxides are :

Oxides	$-\Delta H_f^0$ (1 atm, 298K)kcal/kgmol
CO	29.6×10^3 (C Amorphous)
CO	26.4×10^3
CO_2	97.2×103 (C Amorphous)
CO_2	94.05×10^3
$H_2O(l)$	68.32×10^3
$H_2O(v)$	57.80×10^3
$SO_2(g)$	70.96×10^3
$SO_3(g)$	94.45×10^3

Heats of formation of some hydrocarbons are:

Hydrocarbons	$-\Delta H_f^0$ (1 atm, 298K)kcal/kgmol
CH_4	17.89×10^3
C_2H_2	54.19×10^3
C_2H_4	12.5×10^3
C_2H_6	20.24×10^3
C_3H_8	24.82×10^3

To Note

$$1 \, \text{kgmol} = 22.4 m^3 (1 \, \text{atm}, 0^{\circ}\text{C})$$

$$1 \, \text{kgmol} = 24.45 m^3 (1 \, \text{atm}, 25^{\circ}\text{C})$$

Consider an example of gaseous fuel of the following composition:

$$CH_4 = 4\%$$

$$C_2H_6 = 3\%$$

$$C_3H_8 = 0.5\%$$

$$N_2 \text{ and } CO_2 = \text{Rest}$$

Let us calculate CV of this fuel

In 1 kg mole of gaseous fuel:

	Kg moles
$CH_4 =$	0.94
$C_2H_6 =$	0.03
$C_3H_8 =$	0.005

Heat of combustion of Methane

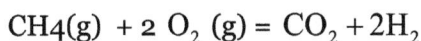

$$CH4(g) + 2 \, O_2 \, (g) = CO_2 + 2H_2$$

Heat of combustion = (Heat of formation of products) − (Heat of formation of reactants)

By substituting the values of heat of formation one obtains heat of combustion of methane equals $194.91 \times 10^3 \text{kcal/kgmol}$.

Similarly, combustion equations for C_2H_6 and C_3H_8 can be written and heat of combustion value can be calculated.

Heat of combustion of $C_2H_6 = -350.56 \times 10^3 \text{kcal/kgmol}$ and of C_3H_8 is

$-498.18 \times 10^3 \text{kcal}/\text{kgmol}$ when reference state of POC is vapour. Thus net calorific value (NCV) of natural gas is

$$|\text{NCV}| = 0.94 \times 194.91 \times 10^3 + 0.03 \times 350.56 \times 10^3 + 0.005 \times 498.18 \times 10^3$$

$$= 196.22 \times 10^3 \text{kcal}/\text{kgmol of natural gas}$$

$$= 8.76 \times 10^3 \text{kcal}/\text{m}^3 \text{ (1 atm and 273K)}$$

Amount of Air

In roasting, air is used for oxidation of sulphides as well as for combustion of coal. Calculation of amount of air is important.

Stoichiometric amount of air (Also termed theoretical air or air for complete conversion of sulphide into oxide or for complete combustion) can be calculated by considering the products of combustion.

Consider the Reaction

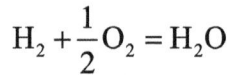

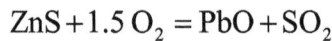

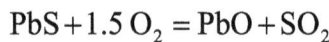

$$C + O_2 = CO_2$$

$$H_2 + \frac{1}{2}O_2 = H_2O$$

$$ZnS + 1.5\,O_2 = PbO + SO_2$$

$$PbS + 1.5\,O_2 = PbO + SO_2$$

The equations are chemically balanced equations. In above equation 1st, we require 1 mole of oxygen to produce 1 moleof CO_2 and in above equation 3rd or 4th we require 1.5 mol of SO_2. It is known that 1 mole of oxygen is obtained from 4.76 moles of air.

Let us calculate stoichiometric amount of air for combustion of solid fuel of composition 84%C, 5%H, 5% moisture and 6% ash, per Kg of coal. Following the stoichiometry of combustion, the amount of air would be

$$\left(\frac{0.84}{12} + \frac{0.05}{2}\right) \times 4.76 = 0.452 \frac{\text{moles}}{\text{kg}}\text{coal}$$

$$= 10.12\text{m}^3 (1\,\text{atm}, 273\text{k})/\text{kgcoal}$$

Note $1\text{kgmole} = 22.4\text{m}^3 (1\,\text{atm}, 273\,\text{k})$

$$\text{Excess air} = \frac{\text{Actual amount of air} - \text{theoretical air}}{\text{theoretical air}} \times 100$$

$$= \frac{\text{actual amount of O}_2 - \text{theoreticalO}_2}{\text{theoreticle air O}_2} \times 100$$

In the above example if actual amount of air is 0.5 mols then

$$\text{Excess air} = \frac{0.5 - 0.452}{0.452} \times 100 = 10.62\%$$

We can also call-that 110.62% theoretical air is used for combustion.

Similarly stoichiometric amount of air can be calculated for above reaction. If reactions occur simultaneously then stoichiometric amount of oxygen is 3 mols and stoichiometric air is 14.28 mols.

Types of Roasting

In dead roasting all sulphides are converted in to oxide. This is done to extract Zn from its sulphide ore.

In sulphatising roasting, sulphide is converted into sulphate as sulphatecan be dissolved easily into an aqueous solution. Typically used for hydro metallurgical extraction of lead sulphide ores.

Thermodynamics of Roasting

Roasting is gas/solid reaction in which sulphide is converted to oxide or sulphate or even to metal. Whether roast product is oxide or sulphate or partially sulphide would depend on temperature and partial pressures.

Phase Rule

Gibbs' phase rule was proposed by Josiah Willard Gibbs in his landmark paper titled *On the Equilibrium of Heterogeneous Substances*, published from 1875 to 1878. The rule applies to non-reactive multi-component heterogeneous systems in thermodynamic equilibrium and is given by the equality

$$F = C - P + 2$$

where F is the number of degrees of freedom, C is the number of components and P is the number of phases in thermodynamic equilibrium with each other.

The number of degrees of freedom is the number of independent intensive variables, i.e. the largest number of thermodynamic parameters such as temperature or pressure that can be varied simultaneously and arbitrarily without affecting one another. An example of one-component system is a system involving one pure chemical, while two-component systems, such as mixtures of water and ethanol, have two chemically independent components, and so on. Typical phases are solids, liquids and gases.

Foundations

- A phase is a form of matter that is homogeneous in chemical composition and physical state. Typical phases are solid, liquid and gas. Two immiscible liquids (or liquid mixtures with different compositions) separated by a distinct boundary are counted as two different phases, as are two immiscible solids.

- The number of components (C) is the number of chemically independent constituents of the system, i.e. the minimum number of independent species necessary to define the composition of all phases of the system.

- The number of degrees of freedom (F) in this context is the number of intensive variables which are independent of each other.

The basis for the rule (Atkins and de Paula, justification 6.1) is that equilibrium between phases places a constraint on the intensive variables. More rigorously, since the phases are in thermodynamic equilibrium with each other, the chemical potentials of the phases must be equal. The number of equality relationships determines the number of degrees of freedom. For example, if the chemical potentials of a liquid and of its vapour depend on temperature (T) and pressure (p), the equality of chemical potentials will mean that each of those variables will be dependent on the other. Mathematically, the equation $\mu_{liq}(T, p) = \mu_{vap}(T, p)$, where μ = chemical potential, defines temperature as a function of pressure or vice versa. (Caution: do not confuse p = pressure with P = number of phases.)

To be more specific, the composition of each phase is determined by $C - 1$ intensive variables (such as mole fractions) in each phase. The total number of variables is $(C - 1)$ $P + 2$, where the extra two are temperature T and pressure p. The number of constraints is $C(P - 1)$, since the chemical potential of each component must be equal in all phases. Subtract the number of constraints from the number of variables to obtain the number of degrees of freedom as $F = (C - 1)P + 2 - C(P - 1) = C - P + 2$.

The rule is valid provided the equilibrium between phases is not influenced by gravitational, electrical or magnetic forces, or by surface area, and only by temperature, pressure, and concentration.

Consequences and Examples

Pure Substances (One Component)

For pure substances $C = 1$ so that $F = 3 - P$. In a single phase ($P = 1$) condition of a pure component system, two variables ($F = 2$), such as temperature and pressure, can be chosen independently to be any pair of values consistent with the phase. However, if the temperature and pressure combination ranges to a point where the pure component undergoes a separation into two phases ($P = 2$), F decreases from 2 to 1. When the system enters the two-phase region, it becomes no longer possible to independently control temperature and pressure.

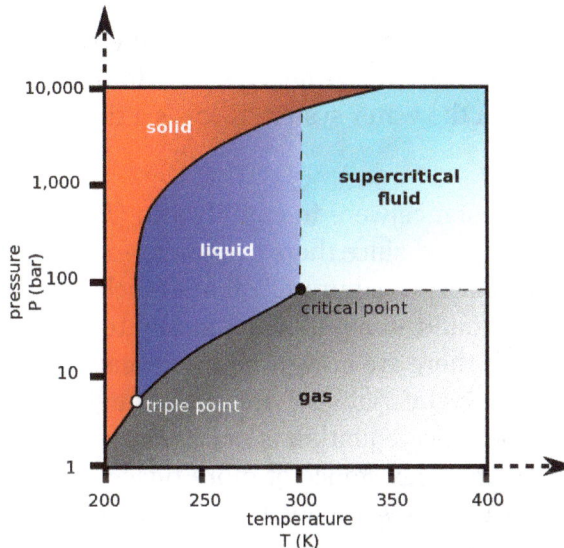

Carbon dioxide pressure-temperature phase diagram showing the triple point and critical point of carbon dioxide

In the phase diagram to the right, the boundary curve between the liquid and gas regions maps the constraint between temperature and pressure when the single-component system has separated into liquid and gas phases at equilibrium. If the pressure is increased by compression, some of the gas condenses and the temperature goes up. If the temperature is decreased by cooling, some of the gas condenses, decreasing the pressure. Throughout both processes, the temperature and pressure stay in the relationship shown by this boundary curve unless one phase is entirely consumed by evaporation or condensation, or unless the critical point is reached. As long as there are two phases, there is only one degree of freedom, which corresponds to the position along the phase boundary curve.

The critical point is the black dot at the end of the liquid–gas boundary. As this point is approached, the liquid and gas phases become progressively more similar until, at the critical point, there is no longer a separation into two phases. Above the critical point and away from the phase boundary curve, $F = 2$ and the temperature and pressure can be controlled independently. Hence there is only one phase, and it has the physical properties of a dense gas, but is also referred to as a supercritical fluid.

Of the other two-boundary curves, one is the solid–liquid boundary or melting point curve which indicates the conditions for equilibrium between these two phases, and the other at lower temperature and pressure is the solid–gas boundary.

Even for a pure substance, it is possible that three phases, such as solid, liquid and vapour, can exist together in equilibrium ($P = 3$). If there is only one component, there are no degrees of freedom ($F = 0$) when there are three phases. Therefore, in a single-component system, this three-phase mixture can only exist at a single temperature and pressure, which is known as a triple point. Here there are two equations $\mu_{sol}(T, p) = \mu_{liq}(T, p) = \mu_{vap}(T, p)$, which are sufficient to determine the two variables T and p. In the diagram for CO_2 the triple point is the point at which the solid, liquid and gas phases come together, at 5.2 bar and 217 K. It is also possible for other sets of phases to form a triple point, for example in the water system there is a triple point where ice I, ice III and liquid can coexist.

If four phases of a pure substance were in equilibrium ($P = 4$), the phase rule would give $F = -1$, which is meaningless, since there cannot be -1 independent variables. This explains the fact that four phases of a pure substance (such as ice I, ice III, liquid water and water vapour) are not found in equilibrium at any temperature and pressure. In terms of chemical potentials there are now three equations, which cannot in general be satisfied by any values of the two variables T and p, although in principle they might be solved in a special case where one equation is mathematically dependent on the other two. In practice, however, the coexistence of more phases than allowed by the phase rule normally means that the phases are not all in true equilibrium.

Two-component Systems

For binary mixtures of two chemically independent components, $C = 2$ so that $F = 4 - P$. In addition to temperature and pressure, the other degree of freedom is the composition of each phase, often expressed as mole fraction or mass fraction of one component.

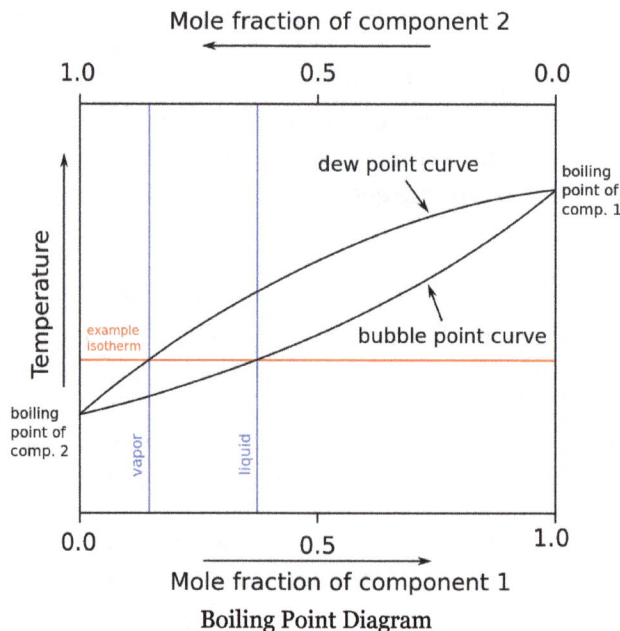

Boiling Point Diagram

As an example, consider the system of two completely miscible liquids such as toluene and benzene, in equilibrium with their vapours. This system may be described by a boiling-point diagram which shows the composition (mole fraction) of the two phases in equilibrium as functions of temperature (at a fixed pressure).

Four thermodynamic variables which may describe the system include temperature (T), pressure (p), mole fraction of component 1 (toluene) in the liquid phase (x_{1L}), and mole fraction of component 1 in the vapour phase (x_{1V}). However since two phases are in equilibrium, only two of these variables can be independent ($F = 2$). This is because the four variables are constrained by two relations: the equality of the chemical potentials of liquid toluene and toluene vapour, and the corresponding equality for benzene.

For given T and p, there will be two phases at equilibrium when the overall composition of the system (system point) lies in between the two curves. A horizontal line (isotherm or tie line) can be drawn through any such system point, and intersects the curve for each phase at its equilibrium composition. The quantity of each phase is given by the lever rule (expressed in the variable corresponding to the x-axis, here mole fraction).

For the analysis of fractional distillation, the two independent variables are instead considered to be liquid-phase composition (x_{1L}) and pressure. In that case the phase rule implies that the equilibrium temperature (boiling point) and vapour-phase composition are determined.

Liquid–vapour phase diagrams for other systems may have azeotropes (maxima or minima) in the composition curves, but the application of the phase rule is unchanged. The only difference is that the compositions of the two phases are equal exactly at the azeotropic composition.

Phase Rule at Constant Pressure

For applications in materials science dealing with phase changes between different solid structures, pressure is often imagined to be constant (for example at one atmosphere), and is ignored as a degree of freedom, so the rule becomes

$$F = C - P + 1.$$

This is sometimes misleadingly called the "condensed phase rule", but it is not applicable to condensed systems which are subject to high pressures (for example, in geology), since the effects of these pressures can be important.

Predominance Area Diagram

Figure shows predominance area diagram for $Ni - S - O$ system, at constant temperature. The phases are shown in the figure.

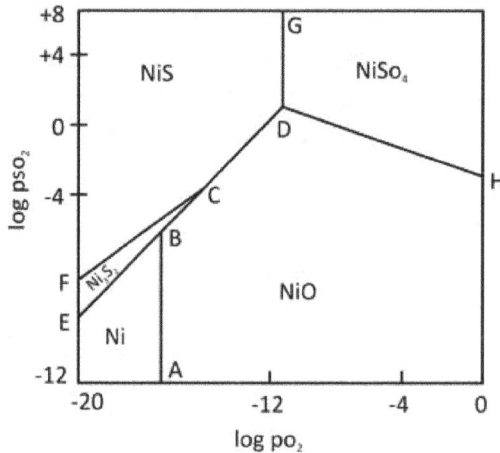

Figure predominance area diagram for $Ni - S - O$ system at constant temperature.

In the figure at points B, C and D; three condensed phases area at equilibrium for a particular value of po_2 and pso_2. Degree of freedom is zero. For example at point B Ni_3S_2 / Ni / NiO can co-exist at fixed po   and pso_2, at point C $NiS - Ni_3S_2 - NiO$ and at po int $D\ NiS - NiSO_4 - NiO\ can\ co - exist$. Thus these points are called invariant point.

The lines describe the equilibrium between any two condensed phases. Along the lines degree of freedom $F = 1$, which means we can vary either po_2 or po_2 to obtain the phases. For example line EB is equilibrium between Ni_3S_2 and Ni, where along line BC equilibrium exists between $Ni_3S_2\ and\ NiO$ Along lines AB and GD equilibrium exists between Ni and NiO, and $NiS\ and\ NiSO_4$. This shows that NiO/Ni or NiS/ $NiSO_4$ equilibrium is independent of pso_2.

The figure also shows predominance areas for a single phase, for example in the area ABCDH NiO is a stable phase, whereas in the area FCDG, NiS is a stable phase. In the area degree of freedom is 2 which means both po_2 and po_2 can be varied to obtain a phase within the area.

Method of Construction

The predominance area diagram depends on the system and temperature. In a two dimensional diagram, temperature is fixed. These are the equilibrium diagrams and hence we have to consider all the phases which can form in a systems.

Consider $Ni - S - O$ system in which $Ni, NiO, Ni_3S_2, NiSO_4$ and NiS phase can form. Let us write chemical equation representing equilibrium between any two condensed phases

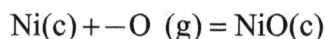

$$Ni(c) + -O\ (g) = NiO(c)$$

$$K_1 = \frac{1}{(po_2)^{0.5}}$$

Since Ni and NiO are pure and hence their activities are unity.

$$LogK_1 = -0.5 \, kgpo_2$$

We see that Ni / NiO equilibrium is independent of Pso_2 and hence it is a vertical line AB in diagram 15.1. The actual values of Pso_2 and Po_2 can be obtained from free energy values.

Similarly line DG represents equilibrium between NiS and $NiSO_4$

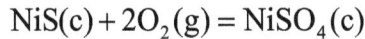

$$NiS(c) + 2O_2(g) = NiSO_4(c)$$

Since activity of condensed phase is unity

$$\log K_2 = -2 \log po_z$$

DG line is also a vertical line.

Consider $NiO - NiSO_4$ equilibrium

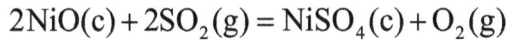

$$2NiO(c) + 2SO_2(g) = NiSO_4(c) + O_2(g)$$

$$K_3 = \frac{po_z}{(pso_z)^2}$$

$$\log p_{so_z} = 0.5 \log po_z - \log K_3$$

We not that equilibrium between NiO and $NiSO_4$ can be attained by varying Pso_2 and Po_2 both and the line DH shows the variation of $\log Pso_2$ against. $\log po_2$.

Similarly

$$Ni_3S_2 + 3.5O_2 = 3 \, NiO + 2SO_2$$

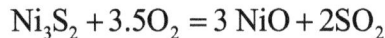

$$\log p_{soz} = 0.5 \log K_2 + 1.75 \log po_z$$

The line BC is the variation between log pSO2 and log po2 for Ni_3S_2 / NiO equilibrium.

The predominance area diagram can be constructed easily by writing $\Delta G°$ values for each reaction.

Utility of predominance – area diagram (PAD)

1. PAD shows the stable phase under different conditions (gas pressures)

2. PAD predicts possible processing routes.

3. One can predict the conditions for formation of a particular phase. In dead roasting of PbS, PbO can form several compounds like $PbSO_4.4PbO$, $PbSO_4$, $2PbO$ and $PbSO_4.PbO$. Dead roasting of PbS is likely to produce PbO and $PbSO_4$.

4. It is possible thermodynamically to produce metal from sulphide by controlling po_z.

Roasting of Complex Sulphide Ores

Additional reactions may occur during the roasting of complex sulphide ore.

Differentsulphide may form solid solutions and even complex sulphides. In iron-copper sulphide ores, numberof ternary phases and also solid solutions of FeS in Cu_2S may form during roasting.

Another phenomenon is the formation of $ZnO - Fe_2O_3$ in roasting of sulphide ores. Since in a complex phase the chemical activity of a given compound is less than the pure compound, its predominance area will expand.

Technology of Roasting

Roasting may be carried out in different furnaces. Multiple hearth furnace was dominant for a long time for roasting of sulphide ores. Now flash roasting is developed. Fluidized bed roasting is also being in use.

Roasting is strongly exothermic process.

A copper concentrate may be roasted autogenously in a multiple hearth furnace provided sulphur is not eliminated completely. Dead roasting would require additional thermal energy. Material balance will also be illustrated first because heat balance cannot be done without materials balance.

Mass Balance

A mass balance, also called a material balance, is an application of conservation of mass to the analysis of physical systems. By accounting for material entering and leaving a system, mass flows can be identified which might have been unknown, or difficult to measure without this technique. The exact conservation law used in the analysis of the system depends on the context of the problem, but all revolve around mass conservation, i.e. that matter cannot disappear or be created spontaneously.

Therefore, mass balances are used widely in engineering and environmental analyses. For example, mass balance theory is used to design chemical reactors, to analyse alternative processes to produce chemicals, as well as to model pollution dispersion and other processes of physical systems. Closely related and complementary analysis techniques include the population balance, energy balance and the somewhat more complex entropy balance. These techniques are required for thorough design and analysis of systems such as the refrigeration cycle.

In environmental monitoring the term budget calculations is used to describe mass balance equations where they are used to evaluate the monitoring data (comparing input and output, etc.) In biology the dynamic energy budget theory for metabolic organisation makes explicit use of mass and energy balances.

Introduction

The general form quoted for a mass balance is *The mass that enters a system must, by conservation of mass, either leave the system or accumulate within the system* .

Mathematically the mass balance for a system without a chemical reaction is as follows:

$$\text{Input} = \text{Output} + \text{Accumulation}$$

Strictly speaking the above equation holds also for systems with chemical reactions if the terms in the balance equation are taken to refer to total mass, i.e. the sum of all the chemical species of the system. In the absence of a chemical reaction the amount of any chemical species flowing in and out will be the same; this gives rise to an equation for each species present in the system. However, if this is not the case then the mass balance equation must be amended to allow for the generation or depletion (consumption) of each chemical species. Some use one term in this equation to account for chemical reactions, which will be negative for depletion and positive for generation. However, the conventional form of this equation is written to account for both a positive generation term (i.e. product of reaction) and a negative consumption term (the reactants used to produce the products). Although overall one term will account for the total balance on the system, if this balance equation is to be applied to an individual species and then the entire process, both terms are necessary. This modified equation can be used not only for reactive systems, but for population balances such as arise in particle mechanics problems. The equation is given below; note that it simplifies to the earlier equation in the case that the generation term is zero.

$$\text{Input} + \text{Generation} = \text{Output} + \text{Accumulation} + \text{Consumption}$$

- In the absence of a nuclear reaction the number of atoms flowing in and out must remain the same, even in the presence of a chemical reaction.

- For a balance to be formed, the boundaries of the system must be clearly defined.

- Mass balances can be taken over physical systems at multiple scales.

- Mass balances can be simplified with the assumption of steady state, in which the accumulation term is zero.

Illustrative Example

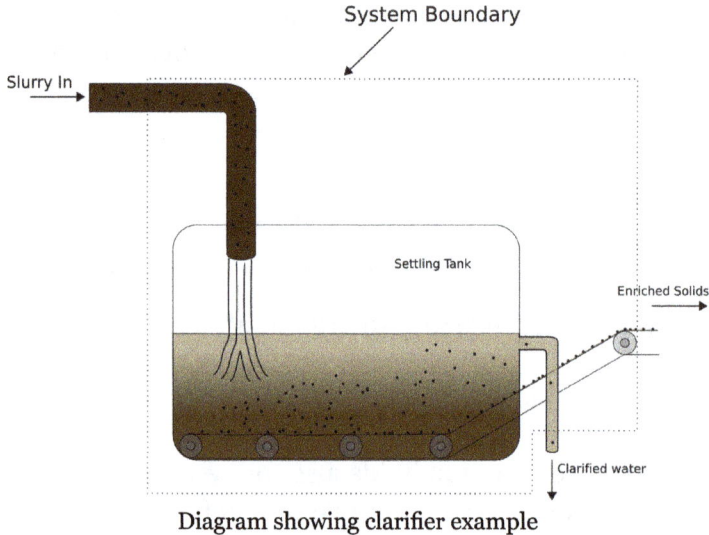

Diagram showing clarifier example

A simple example can illustrate the concept. Consider the situation in which a slurry is flowing into a settling tank to remove the solids in the tank. Solids are collected at the bottom by means of a conveyor belt partially submerged in the tank, and water exits via an overflow outlet.

In this example, there are two substances: solids and water. The water overflow outlet carries an increased concentration of water relative to solids, as compared to the slurry inlet, and the exit of the conveyor belt carries an increased concentration of solids relative to water.

Assumptions

- Steady state

- Non-reactive system

Analysis

Suppose that the slurry inlet composition (by mass) is 50% solid and 50% water, with a mass flow of 100 kg/min. The tank is assumed to be operating at steady state, and as such accumulation is zero, so input and output must be equal for both the solids and water. If we know that the removal efficiency for the slurry tank is 60%, then the water outlet will contain 20 kg/min of solids (40% times 100 kg/min times 50% solids). If we measure the flow rate of the combined solids and water, and the water outlet is shown to be 65 kg/min, then the amount of water exiting via the conveyor belt must be 10 kg/min. This allows us to completely determine how the mass has been distributed in the system with only limited information and using the mass balance relations across the system boundaries.

Mass Feedback (Recycle)

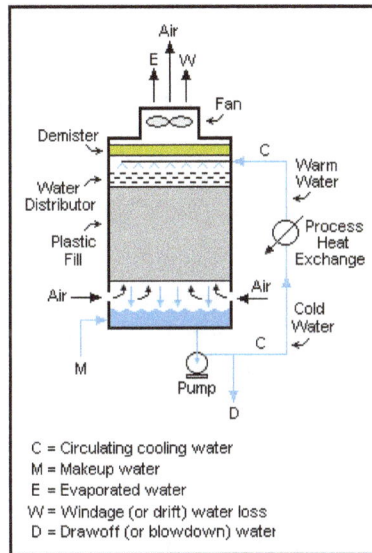

Cooling towers are a good example of a recycle system

Mass balances can be performed across systems which have cyclic flows. In these systems output streams are fed back into the input of a unit, often for further reprocessing.

Such systems are common in grinding circuits, where grain is crushed then sieved to only allow fine particles out of the circuit and the larger particles are returned to the roller mill (grinder). However, recycle flows are by no means restricted to solid mechanics operations; they are used in liquid and gas flows, as well. One such example is in cooling towers, where water is pumped through a tower many times, with only a small quantity of water drawn off at each pass (to prevent solids build up) until it has either evaporated or exited with the drawn off water.

The use of the recycle aids in increasing overall conversion of input products, which is useful for low per-pass conversion processes (such as the Haber process).

Differential Mass Balances

A mass balance can also be taken differentially. The concept is the same as for a large mass balance, but it is performed in the context of a limiting system (for example, one can consider the limiting case in time or, more commonly, volume). A differential mass balance is used to generate differential equations that can provide an effective tool for modelling and understanding the target system.

The differential mass balance is usually solved in two steps: first, a set of governing differential equations must be obtained, and then these equations must be solved, either analytically or, for less tractable problems, numerically.

The following systems are good examples of the applications of the differential mass balance:

1. Ideal (stirred) Batch reactor

2. Ideal tank reactor, also named Continuous Stirred Tank Reactor (CSTR)

3. Ideal Plug Flow Reactor (PFR)

Ideal Batch Reactor

The ideal completely mixed batch reactor is a closed system. Isothermal conditions are assumed, and mixing prevents concentration gradients as reactant concentrations decrease and product concentrations increase over time. Many chemistry textbooks implicitly assume that the studied system can be described as a batch reactor when they write about reaction kinetics and chemical equilibrium. The mass balance for a substance A becomes

$$IN + PROD = OUT + ACC$$

$$0 + r_A V = 0 + \frac{dn_A}{dt}$$

where r_A denotes the rate at which substance A is produced, V is the volume (which may be constant or not), n_A the number of moles (n) of substance A.

In a fed-batch reactor some reactants/ingredients are added continuously or in pulses (compare making porridge by either first blending all ingredients and then letting it boil, which can be described as a batch reactor, or by first mixing only water and salt and making that boil before the other ingredients are added, which can be described as a fed-batch reactor). Mass balances for fed-batch reactors become a bit more complicated.

Reactive Example

In the first example, we will show how to use a mass balance to derive a relationship between the percent excess air for the combustion of a hydrocarbon-base fuel oil and the percent oxygen in the combustion product gas. First, normal dry air contains

0.2095 mol of oxygen per mole of air, so there is one mole of O_2 in 4.773 mol of dry air. For stoichiometric combustion, the relationships between the mass of air and the mass of each combustible element in a fuel oil are:

$$\text{Carbon: } \frac{\text{mass of air}}{\text{mass of C}} = \frac{4.773 \times 28.96}{12.01} = 11.51$$

$$\text{Hydrogen: } \frac{\text{mass of air}}{\text{mass of H}} = \frac{\frac{1}{4}(4.773) \times 28.96}{1.008} = 34.28$$

$$\text{Sulfur: } \frac{\text{mass of air}}{\text{mass of S}} = \frac{4.773 \times 28.96}{32.06} = 4.31$$

Considering the accuracy of typical analytical procedures, an equation for the mass of air per mass of fuel at stoichiometric combustion is:

$$\frac{\text{mass of air}}{\text{mass of fuel}} = AFR_{mass} = 11.5(wC) + 34.3(wH) + (wS - wO)$$

where wC, wH, wS, and wO refer to the mass fraction of each element in the fuel oil, sulfur burning to SO2, and AFR_{mass} refers to the air-fuel ratio in mass units.

For 1 kg of fuel oil containing 86.1% C, 13.6% H, 0.2% O, and 0.1% S the stoichiometric mass of air is 14.56 kg, so AFR = 14.56. The combustion product mass is then 15.56 kg. At exact stoichiometry, O_2 should be absent. At 15 percent excess air, the AFR = 16.75, and the mass of the combustion product gas is 17.75 kg, which contains 0.505 kg of excess oxygen. The combustion gas thus contains 2.84 percent O_2 by mass. The relationships between percent excess air and $\%O_2$ in the combustion gas are accurately expressed by quadratic equations, valid over the range 0–30 percent excess air:

$$\% \text{ excess air} = 1.2804 \times (\%O_2 \text{ in combustion gas})^2 + 4.49 \times (\%O_2 \text{ in combustion gas})$$

$$\%O_2 \text{ in combustion gas} = -0.00138 \times (\% \text{ excess air})^2 + 0.210 \times (\% \text{ excess air})$$

In the second example we will use the law of mass action to derive the expression for a chemical equilibrium constant.

Assume we have a closed reactor in which the following liquid phase reversible reaction occurs:

$$aA + bB \leftrightarrow cC + dD$$

The mass balance for substance A becomes

$$IN + PROD = OUT + ACC$$

$$0 + r_A V = 0 + \frac{dn_A}{dt}$$

As we have a liquid phase reaction we can (usually) assume a constant volume and since $n_A = V * C_A$ we get

$$r_A V = V \frac{dC_A}{dt}$$

or

$$r_A = \frac{dC_A}{dt}$$

In many textbooks this is given as the definition of reaction rate without specifying the implicit assumption that we are talking about reaction rate in a closed system with only one reaction. This is an unfortunate mistake that has confused many students over the years.

According to the law of mass action the forward reaction rate can be written as

$$r_1 = k_1 [A]^a [B]^b$$

and the backward reaction rate as

$$r_{-1} = k_{-1} 1 [C]^c [D]^d$$

The rate at which substance A is produced is thus

$$r_A = a(r_{-1} - r_1)$$

and since, at equilibrium, the concentration of A is constant we get

$$r_A = a(r_{-1} - r_1) = \frac{dC_A}{dt} = 0$$

or, rearranged

$$\frac{k_1}{k_{-1}} = \frac{[C]^c [D]^d}{[A]^a [B]^b} = K_{eq}$$

Ideal Tank Reactor/continuously Stirred Tank Reactor

The continuously mixed tank reactor is an open system with an influent stream of reactants and an effluent stream of products. A lake can be regarded as a tank reactor, and lakes with long turnover times (e.g. with low flux-to-volume ratios) can for many purposes be regarded as continuously stirred (e.g. homogeneous in all respects). The mass balance then becomes

$$IN + PROD = OUT + ACC$$

$$Q_0 \cdot C_{A,0} + r_A \cdot V = Q \cdot C_A + \frac{dn_A}{dt}$$

where Q_0 and Q denote the volumetric flow in and out of the system respectively and $C_{A,0}$ and C_A the concentration of A in the inflow and outflow respective. In an open system we can never reach a chemical equilibrium. We can, however, reach a steady state where all state variables (temperature, concentrations etc.) remain constant ($ACC = 0$).

Example

Consider a bathtub in which there is some bathing salt dissolved. We now fill in more water, keeping the bottom plug in. What happens?

Since there is no reaction, $PROD = 0$ and since there is no outflow $Q = 0$. The mass balance becomes

$$IN + PROD = OUT + ACC$$

$$Q_0 \cdot C_{A,0} + 0 = 0 \cdot C_A + \frac{dn_A}{dt}$$

or

$$Q_0 \cdot C_{A,0} = \frac{dC_A V}{dt} = V\frac{dC_A}{dt} + C_A\frac{dV}{dt}$$

Using a mass balance for total volume, however, it is evident that $\frac{dV}{dt} = Q_0$ and that $V = V_{t=0} + Q_0 t$. Thus we get

$$\frac{dC_A}{dt} = \frac{Q_0}{(V_{t=0} + Q_0 t)}\left(C_{A,0} - C_A\right)$$

Note that there is no reaction and hence no reaction rate or rate law involved, and yet $\frac{dC_A}{dt} \neq 0$. We can thus draw the conclusion that reaction rate can not be defined in a general manner using $\frac{dC}{dt}$. One must first write down a mass balance before a link between $\frac{dC}{dt}$ and the reaction rate can be found. Many textbooks, however, define reaction rate as

$$v = \frac{dC_A}{dt}$$

without mentioning that this definition implicitly assumes that the system is closed, has a constant volume and that there is only one reaction.

Ideal Plug Flow Reactor (PFR)

The idealized plug flow reactor is an open system resembling a tube with no mixing in the direction of flow but perfect mixing perpendicular to the direction of flow. Often used for systems like rivers and water pipes if the flow is turbulent. When a mass balance is made for a tube, one first considers an infinitesimal part of the tube and make a mass balance over that using the ideal tank reactor model. That mass balance is then integrated over the entire reactor volume to obtain:

$$\frac{d(Q \cdot C_A)}{dV} = r_A$$

In numeric solutions, e.g. when using computers, the ideal tube is often translated to a series of tank reactors, as it can be shown that a PFR is equivalent to an infinite number of stirred tanks in series, but the latter is often easier to analyze, especially at steady state.

More Complex Problems

In reality, reactors are often non-ideal, in which combinations of the reactor models above are used to describe the system. Not only chemical reaction rates, but also mass transfer rates may be important in the mathematical description of a system, especially in heterogeneous systems.

As the chemical reaction rate depends on temperature it is often necessary to make both an energy balance (often a heat balance rather than a full-fledged energy balance) as well as mass balances to fully describe the system. A different reactor model might be needed for the energy balance: A system that is closed with respect to mass might be open with respect to energy e.g. since heat may enter the system through conduction.

Commercial Use

In industrial process plants, using the fact that the mass entering and leaving any portion of a process plant must balance, data validation and reconciliation algorithms may be employed to correct measured flows, provided that enough redundancy of flow measurements exist to permit statistical reconciliation and exclusion of detectably erroneous measurements. Since all real world measured values contain inherent error, the reconciled measurements provide a better basis than the measured values do for financial reporting, optimization, and regulatory reporting. Software packages exist to make this commercially feasible on a daily basis.

Calculation Procedure

1. SiO_2, Al_2O_3 .and CaO of ore concentrate enter into roast product.

2. Fe is oxidized to Fe_2O_3 and Fe_2O_4 depending on oxidizing condition.

3. In partial roasting, usually not all S of concentrate is oxidized.

4. Sulphur in the roast product is present as Cu_2S and FeS.

5. In the roast product oxygen is in combined form either with Fe or with Cu.

6. Theoretical amount of air can always be calculated once the reactants and products are specified. Here balanced chemical equation helps very often to calculate theoretical air.

7. 1 mole of air contains 0.21mols of O_2 and 0.79mols of N_2 .1 mol of oxygen and 3.76mols of N_2 forms 4.76mols.

To find weight of roasted ore one can do either SiO_2 balance or Al_2O_3 balance or even CaO balance.

SiO_2 balance and Al_2O_3 balance give weight of roasted product 791kg .

Roasted ore contains Cu_2S, FeS, Fe_2O_3, Fe_3O_3, CaO SiO_2 and Al_2O_3.

Amount of CaO, SiO_2 and Al_2O_3 can be determined from their percentages in roast product, these are 3.2kg, 190kg and 53kg respectively.

All Cu in roasted ore is present as Cu_2S ,therefore amount of $Cu_2S = 109.75kg$.

It is not known in what form Fe is present in the roasted product. Problem says Fe is present as Fe_2O_3 and Fe_3O_4. It is important to perform S balance.

Sulphur in roasted ore $= 79.1kg$.

S in Cu_2S + S in $FeS = 79.1$

From the amount of Cu_2S we can find S in Cu_2S and we can obtain S in FeS which is equal to 57.15kg

Amount of $FeS = 157.16kg$

Let x Kg Fe_2O_3 and y Kg Fe_3O_4 in roasted ore.

Fe oxidized to Fe_2O_3 and Fe_3O_4 = Total Fe – Fe in Fes $= 197Kg$.

O_2 in roasted ore is either with Fe_2O_3 or Fe_3O_4.

Performing Fe balance and oxygen balance

$$0.7x + 0.72y = 197$$

$$0.3x + 0.28y = 81$$

$$x = 158kg \text{ and } y = 120kg.$$

Proximate analysis of roast product is

Cu_2S	109.75
Al_2O_3	53.00
SiO_2	190.00
CaO	3.20
FeS	157.16
Fe_2O_3	158.00
Fe_3O_4	120.00
Total	**791.11**

Volume and percentage composition of gases

$$\text{Total S oxidized} = 9.059 \text{ kg mols}$$

$$\text{S oxidized to } So_2 = 7.70 \text{ kg mols}$$

$$\text{S oxidized to } SO_3 = 1.36 \text{kg mols}$$

Since following oxidation reactions are occurring:

$$S + O_2 = SO_2$$

$$S + 1.5 \, O_2 = SO_3$$

$$2Fe + 1.5 \, O_2 = Fe_2O_3$$

$$3Fe + 2O_2 = Fe_3O_4$$

$$C + O_2 = CO_2$$

Form the amounts of elements oxidized

$$\text{Stoichiometric oxygen} = 12.667 \text{ kg mols}$$

$$\text{Actual O}_2 \text{ supplied} = \frac{5000 \times 0.21}{22.4} = 46.875 \text{kg mols}$$

$$\text{Actual N}_2 \text{ supplied} = 176.34 \text{ kg mols}$$

$$\text{excess O}_2 = 34.21 \text{ kg mols}$$

Composition	Kg mols	%
SO_2	7.70	3.50
SO_3	1.36	0.63
O_2	34.21	15.54
N_2	176.34	80.14
CO_2	0.42	0.19

$$\text{Volume of gases} = 4928.67 \text{m}^3$$

$$\text{excess air} = \frac{\text{excess O}_2}{\text{Theoretical O}_2} \times 100$$

$$= 273.7\%$$

Heat Balance in Roasting

Heat balanceis an important exercise in all high temperature processes. Heat balance can be used to determine the temperature, theoretically, which can be attained during roasting. Knowledge of temperature is useful to take a decision about the refractory-material

Heat Balance

At unsteady state

Heat input=Heat output +Heat accumulation +Heat losses

Inhightemperature processes, certain amount of thermal energy is retained within the reactor to maintainprocessing temperature. At steady state and at constant tempera-ture.

Heat input=Heat output+ Heat losses

Calculation Procedurefor Temperature

In roastingore concentrate and coal are mixed to convert sulphide into oxide. Heat generated by roasting and combustion raises the temperature of the products of roasting and combustion. Thermodynamically it is possible to calculate the temperature under adiabatic condition,i.e. no heatis allowed to loss. This calculated temperature can then be adjusted by incorporating heat losses. We consider an adiabaticprocess of roasting:

Various Heat Input Terms are:

- Heat of reaction

- Heat of combustion

- Sensible heat in reactants

This amount of heat input raises the temperature of products.

The products consists of solid products and gaseous

$$\Sigma_{i=1}^{n}(H_T - H_{298})_i = \Sigma_{i=1}^{n} m_i c_{pi} \int_{298}^{T} Cp(dT)$$

Here T is the adiabatic temperature. Thus for adiabatic roasting and combustion.

$\Sigma(H_T - H_{298})$ for reactants +Heat of reaction and combustion

$$= \Sigma_{i-1}^{n} m_i c_{pi} \int_{298}^{T} Cp(dT)$$

Where i is number of reactants and combustion products.

In equation 3 heat of reaction & combustion has to be added, since it is liberated.

If the material balance for roasting is known, as well as temperatures of fuel and air, heat content data and heat of reaction data can be used to setup an equation in which temperature of the products is unknown. Let me illustrate it with a problem.

References

- Dowling, A. P. (2000a). "Vortices, sound and flame – a damaging combination". The Aeronautical Journal of the RaeS

- Greenwood, Norman N.; Earnshaw, Alan (1997). Chemistry of the Elements (2nd ed.). Butterworth-Heinemann. ISBN 0-08-037941-9

- Himmelblau, David M. (1967). Basic Principles and Calculations in Chemical Engineering (2nd ed.). Prentice Hall

- Ray, H.S.; et al. (1985). Extraction of Nonferrous Metals. Affiliated East-West Press Private Limited. pp. 131, 132. ISBN 81-85095-63-9

- Shuttle-Mir History/Science/Microgravity/Candle Flame in Microgravity (CFM) – MGBX. Spaceflight.nasa.gov (1999-07-16). Retrieved on 2010-09-28

- Atkins, P.W.; de Paula, J. (2006). Physical chemistry (8th ed.). Oxford University Press. ISBN 0-19-870072-5. Chapter 6

- Weber, Walter J., Jr. (1972). Physicochemical Processes for Water Quality Control. Wiley-Interscience. ISBN 0-471-92435-0

- Perry, Robert H.; Chilton, Cecil H.; Kirkpatrick, Sidney D. (1963). Chemical Engineers' Handbook (4th ed.). McGraw-Hill. pp. 4–21

Smelting: An Overview

Smelting, a part of extraction metallurgy, is used to produce base metals from ores. Smelting is combined with reduction for the extraction process. The reducing agent removes impurities and leaves behind the designated base metal. The section also lays impetus on iron-making and delves into the study of different smelting processes. The aspects elucidated in this chapter are of vital importance, and provide a better understanding of metallurgy.

Smelting

Electric phosphate smelting furnace in a TVAchemical plant

Smelting is a form of extractive metallurgy; its main use is to produce a base metal from its ore. This includes production of silver, iron, copper and other base metals from their ores. Smelting makes use of heat and a chemical reducing agent to decompose the ore, driving off other elements as gases or slag and leaving just the metal base behind. The reducing agent is commonly a source of carbon such as coke, or in earlier times charcoal.

The carbon (or carbon monoxide derived from it) removes oxygen from the ore, leaving behind the elemental metal. The carbon is thus oxidized in two stages, producing first carbon monoxide and then carbon dioxide. As most ores are impure, it is often necessary to use flux, such as limestone, to remove the accompanying rock gangue as slag.

Plants for the electrolytic reduction of aluminium are also generally referred to as aluminium smelters.

Process

Smelting involves more than just melting the metal out of its ore. Most ores are a chemical compound of the metal with other elements, such as oxygen (as an oxide), sulfur (as a sulfide) or carbon and oxygen together (as a carbonate). To produce the metal, these compounds have to undergo a chemical reaction. Smelting therefore consists of using suitable reducing substances that will combine with those oxidizing elements to free the metal.

Roasting

In the case of carbonates and sulfides, a process called "roasting" drives out the unwanted carbon or sulfur, leaving an oxide, which can be directly reduced. Roasting is usually carried out in an oxidizing environment. A few practical examples:

- Malachite, a common ore of copper, is primarily copper carbonate hydroxide $Cu_2(CO_3)(OH)_2$. This mineral undergoes thermal decomposition to $2CuO$, CO_2, and H_2O in several stages between 250 °C and 350 °C. The carbon dioxide and water are expelled into the atmosphere, leaving copper(II) oxide which can be directly reduced to copper.

- Galena, the most common mineral of lead, is primarily lead sulfide (PbS). The sulfide is oxidized to a sulfite ($PbSO_3$) which thermally decomposes into lead oxide and sulfur dioxide gas. (PbO and SO_2) The sulfur dioxide is expelled (like the carbon dioxide in the previous example), and the lead oxide is reduced as below.

Reduction

Reduction is the final, high-temperature step in smelting. It is here that the oxide becomes the elemental metal. A reducing environment (often provided by carbon monoxide, made by incomplete combustion, produced in an air-starved furnace) pulls the final oxygen atoms from the raw metal. The required temperature varies over a very large range, both in absolute terms and in terms of the melting point of the base metal. A few examples:

- iron oxide becomes metallic iron at roughly 1250 °C (2282 °F or 1523.15 K), almost 300 degrees *below* iron's melting point of 1538 °C (2800.4 °F or 1811.15 K)

- mercuric oxide becomes vaporous mercury near 550 °C (1022 °F or 823.15 K), almost 600 degrees *above* mercury's melting point of -38 °C (-36.4 °F or 235.15 K)

Flux and slag can provide a secondary service after the reduction step is complete: They provide a molten cover on the purified metal, preventing it from coming into contact with oxygen while it is still hot enough to oxidize readily.

Fluxes

Fluxes are used in smelting for several purposes, chief among them catalyzing the desired reactions and chemically binding to unwanted impurities or reaction products. Calcium oxide, in the form of lime, was often used for this purpose, since it could react with the carbon dioxide and sulfur dioxide produced during roasting and smelting to keep them out of the working environment.

History

Of the seven metals known in antiquity, only gold occurred regularly in native form in the natural environment. The others – copper, lead, silver, tin, iron and mercury – occur primarily as minerals, though copper is occasionally found in its native state in commercially significant quantities. These minerals are primarily carbonates, sulfides, or oxides of the metal, mixed with other components such as silica and alumina. Roasting the carbonate and sulfide minerals in air converts them to oxides. The oxides, in turn, are smelted into the metal. Carbon monoxide was (and is) the reducing agent of choice for smelting. It is easily produced during the heating process, and as a gas comes into intimate contact with the ore.

In the Old World, humans learned to smelt metals in prehistoric times, more than 8000 years ago. The discovery and use of the "useful" metals — copper and bronze at first, then iron a few millennia later — had an enormous impact on human society. The impact was so pervasive that scholars traditionally divide ancient history into Stone Age, Bronze Age, and Iron Age.

In the Americas, pre-Inca civilizations of the central Andes in Peru had mastered the smelting of copper and silver at least six centuries before the first Europeans arrived in the 16th century, while never mastering the smelting of metals such as iron for use with weapon-craft.

Tin and Lead

In the Old World, the first metals smelted were tin and lead. The earliest known cast lead beads were found in the Çatal Höyük site in Anatolia (Turkey), and dated from about 6500 BC, but the metal may have been known earlier.

Since the discovery happened several millennia before the invention of writing, there is no written record about how it was made. However, tin and lead can be smelted by placing the ores in a wood fire, leaving the possibility that the discovery may have occurred by accident.

Although lead is a common metal, its discovery had relatively little impact in the ancient world. It is too soft to be used for structural elements or weapons, except for the fact that it is exceptionally dense, making it ideal for sling projectiles. However, being easy to cast and shape, it came to be extensively used in the classical world of Ancient Greece and Ancient Rome for piping and storage of water. It was also used as a mortar in stone buildings.

Tin was much less common than lead and is only marginally harder, and had even less impact by itself.

Copper and Bronze

After tin and lead, the next metal to be smelted appears to have been copper. How the discovery came about is a matter of much debate. Campfires are about 200 °C short of the temperature needed for that, so it has been conjectured that the first smelting of copper may have been achieved in pottery kilns. The development of copper smelting in the Andes, which is believed to have occurred independently of that in the Old World, may have occurred in the same way. The earliest current evidence of copper smelting, dating from between 5500 BC and 5000 BC, has been found in Pločnik and Belovode, Serbia. A mace head found in Can Hasan, Turkey and dated to 5000 BC, once thought to be the oldest evidence, now appears to be hammered native copper.

By combining copper with tin and/or arsenic in the right proportions one obtains bronze, an alloy which is significantly harder than copper. The first copper/arsenic bronzes date from 4200 BC from Asia Minor. The Inca bronze alloys were also of this type. Arsenic is often an impurity in copper ores, so the discovery could have been made by accident; but eventually arsenic-bearing minerals were intentionally added during smelting.

Copper–tin bronzes, harder and more durable, were developed around 3200 BC, also in Asia Minor.

The process through which the smiths learned to produce copper/tin bronzes is once again a mystery. The first such bronzes were probably a lucky accident from tin contamination of copper ores, but by 2000 BC, we know that tin was being mined on purpose for the production of bronze. This is amazing, given that tin is a semi-rare metal, and even a rich cassiterite ore only has 5% tin. Also, it takes special skills (or special instruments) to find it and to locate the richer lodes. But, whatever steps were taken to learn about tin, these were fully understood by 2000 BC.

The discovery of copper and bronze manufacture had a significant impact on the history of the Old World. Metals were hard enough to make weapons that were heavier, stronger, and more resistant to impact-related damage than their wood, bone, or stone equivalents. For several millennia, bronze was the material of choice for weapons such as swords, daggers, battle axes, and spear and arrow points, as well as protective gear such as shields, helmets, greaves (metal shin guards), and other body armor. Bronze also supplanted stone, wood, and organic materials in all sorts of tools and household utensils, such as chisels, saws, adzes, nails, blade shears, knives, sewing needles and pins, jugs, cooking pots and cauldrons, mirrors, horse harnesses, and much more. Tin and copper also contributed to the establishment of trade networks spanning large areas of Europe and Asia, and had a major effect on the distribution of wealth among individuals and nations.

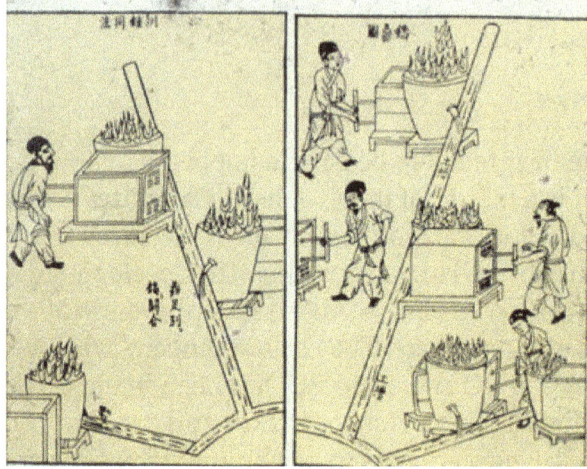

Casting bronze ding-tripods, from the Chinese *Tiangong Kaiwu* encyclopedia of Song Yingxing

Early Iron Smelting

Where and how iron smelting was discovered is widely debated, and remains uncertain due to the significant lack of production finds. Nevertheless, there is some consensus that iron technology originated in the Near East, perhaps in Eastern Anatolia.

In Ancient Egypt, somewhere between the Third Intermediate Period and 23rd Dynasty (ca. 1100–750 BC), there are indications of iron working. Significantly though, no evidence for the smelting of iron from ore has been attested to Egypt in any (pre-modern) period. There is a further possibility of iron smelting and working in West Africa by 1200 BC. In addition, very early instances of carbon steel were found to be in production around 2000 years before the present in northwest Tanzania, based on complex preheating principles. These discoveries are significant for the history of metallurgy.

Most early processes in Europe and Africa involved smelting iron ore in a bloomery, where the temperature is kept low enough so that the iron does not melt. This produces a spongy mass of iron called a bloom, which then has to be consolidated with a hammer. The earliest evidence to date for the bloomery smelting of iron is found at Tell Hammeh, Jordan, and dates to 930 BC.

Later Iron Smelting

From the medieval period, the process of direct reduction in bloomeries began to be replaced by an indirect process. In this, a blast furnace was used to make pig iron, which then had to undergo a further process to make forgeable bar iron. Processes for the second stage include fining in a finery forge and, from the Industrial Revolution, puddling. However both processes are now obsolete, and wrought iron is now hardly made. Instead, mild steel is produced from a bessemer converter or by other means including smelting reduction processes such as the Corex Process.

Base Metals

Cowles Syndicate of Ohio in Stoke-upon-TrentEngland, late 1880s. British Aluminium used the process of Paul Héroult about this time.

The ores of base metals are often sulfides. In recent centuries, reverberatory furnaces have been used. These keep the fuel and the charge being smelted separate. Traditionally these were used for carrying out the first step: formation of two liquids, one an oxide slag containing most of the impurity elements, and the other a sulfide matte containing the valuable metal sulfide and some impurities. Such "reverb" furnaces are today about 40 m long, 3 m high and 10 m wide. Fuel is burned at one end and the heat melts the dry sulfide concentrates (usually after partial roasting), which are fed through the openings in the roof of the furnace. The slag floats on top of the heavier matte, and is removed and discarded or recycled. The sulfide matte is then sent to the converter. The precise details of the process will vary from one furnace to another depending on the mineralogy of the orebody from which the concentrate originates.

While reverberatory furnaces were very good at producing slags containing very little copper, they were relatively energy inefficient and produced a low concentration of sulfur dioxide in their off-gases that made it difficult to capture, and consequently, they have been supplanted by a new generation of copper smelting technologies. More recent furnaces have been designed based upon bath smelting, top jetting lance smelting, flash smelting and blast furnaces. Some examples of bath smelters include the Noranda furnace, the Isasmelt furnace, the Teniente reactor, the Vunyukov smelter and the SKS technology to name a few. Top jetting lance smelters include the Mitsubishi smelting reactor. Flash smelters account for over 50% of the world's copper smelters. There are many more varieties of smelting processes, including the Kivset, Ausmelt, Tamano, EAF, and BF.

Salient Features of Lead Reduction Smelting

A salient feature of lead blast furnace will be given in below.

- Pb is produced by carbon reduction of the sinter containing lead oxide in a lead blast furnace. Sinter consists of $PbS, PbO, PbSO_4, SiO_2, Al_2O_3$ etc

- Height of blast furnace 8m, top dia 3/4m, crucible depth. and stack height 5 m.

- Temperature at bosh $N1200°c$.

- Air is blown in through 15 tuyeres(15-20) around the bosh to oxidize C charged with sinter which produces required heat.

- At top of furnace bag houses are provided to collect lead furnaces from outgoing gases.

- Limestone and quartz are added to make a slag.

Scrap iron is charged. The following reactions occurs,

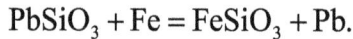

$$PbSiO_3 + Fe = FeSiO_3 + Pb.$$

$$PbO + Fe = FeO + Pb.$$

The products of smelting are liquid lead, matte, speiss, slag and gases. Lead has density 11 g/cm3, matte containing copper has density 5 g/cm3, whereas speiss has density 6 g/cm3

Material Balance in Lead Smelting

A) Lead ore concentrate of composition 60%Pbs, 4% FeS_2, 3% Al_2O_3, 2% C_aO and 31% SiO_2 is roasted by using 300% theoretical air. During roasted all FeS_2, and 90%PbS is oxidized. Determine for 1000kg concentrate.

1) Volume of air at 1 atm and 273 K

2) Amount and composition of roasted product

3) Amount and composition of flue gas

B) The roasted product determined above is smelted in blast furnace using flux as Fe_2O_3 and $CaCO_3$ to produce slag having a ratio $SiO_2 : FeO : CaO$ equal to 35:50:15. The amount of coke is 16% of weight of concentrate. The coke contains 85% C and 15% SiO_2

Find:

4) Amount of each flux

5) Amount of slag

Solution

Molecular weight $Pb = 239, Fe = 56, Al = 27, Ca = 40, Si = 28$

Roasting reactions are

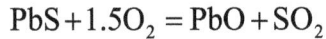

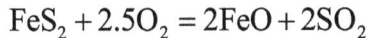

$$PbS + 1.5O_2 = PbO + SO_2$$

$$FeS_2 + 2.5O_2 = 2FeO + 2SO_2$$

Volume of Air = (Theoretical Amount)x3

Theoretical air can be calculated from the stoichiometric equation.

Volume of air = $135m^3$

Roasted product consists of $PbS, PbO, FeO, Al_2O_3, CaO$ and SiO_2. We have done earlier materials balance

Amount of roasted product =948 kg

Amount of flue gas=59 kg mole

Smelting of roast product constitutes part B. Flux used is limestone and Fe_2O_3

Amount of $CaCO_3$ (det er min ed from CaO) = 220Kg.

Amount of $Fe_2O_{3,}$ = 503 Kg.

Amount of slag =984.3 kg

The charge for Pb blast furnace consists of Roasted Ore *PbO* 31%, *PbS* 19%, *SiO_2* 40% *and FeO* 10% (1000 Kg)

Pyritecinder $Fe_2O_3, 88\%, CuS_5 5\%, SiO_2 7\%$

(600Kg)

Coke C 89%, $SiO_2 11\%$

(180Kg)

Flux

$CaCO_3 100\%$

(220 Kg)

Matte contains all of . Cu, S and 2% Pb No Pb is lost in slag or gases.

The gases contain $CO : CO_2 = 1.1$ by volume

Calculate:

a) Charge balance of furnace

b) Proximate composition of matte and slag

c) Volume of blast used (at 1 atm and 273K)

Technique of smelting and converting as applied for copper production is not applicable for production of Zn from sulphide ore since ZnS does not melt even at $1500°C$. ZnS Concentrate is sinter roasted, that means sintering and roasting of sulphide concentrate occur simultaneously.

The burden Zinc sinter and coke is preheated to about $800°C$ before it is charged into Imperial Smelting Furnace (here after ISF), which is very similar to the blast furnace. Preheated air is introduced at the tuyeres that reacts with coke to give CO and generates a temperature of about $1300°C$. Rising CO within the shaft reacts with ZnO to produce a gas mixture of about 5%Zn, 10%CO_2 and 20%CO and N_2.

The hot gas mixture is shock cooled in a lead bath to about $500°C$. Zinc condenses and dissolves in lead without reoxidation.

Materials and Heat in Balance Imperial Smelting Process

In an imperial smelting blast furnace the sinter consists of 50% ZnO, 20%PbO, 20%FeO and 10%SiO_2

The coke is regarded pure amorphous carbon. All ZnO is reduced to Zn vapor and PbO to liquid lead, whereas FeO and SiO_2 form a molten slag. The gas composition after reduction is 7%Zn with CO_2 / CO ratio 0.5

1. Make a mass balance and calculate wt. of carbon and volume of air and volume composition of top gas.

2. Make an overall heat balance when the raw materials are introduced at 1100K, liquid products are withdrawn at 1600K, and the top gas temperature is 1300K.

Materials Balance

Basis: 1000 kg sinter.

The reduction equations for ZnO and PbO can be written by using the moles of ZnO and $CO : CO_2$ participating in the reaction and taking into account $CO : CO_2$ ratio in the exit gas is 2:1. This approach will directly give the moles of products. However, other approaches may also be used

	Kg moles
Zn(v) =	6.173
Pb(l) =	0.897
CO_2 =	7.070
CO =	14.140
N_2 =	39.875

The exit gas amount as determined from Zn vapour balance is 88.186 kg moles.

The amount of exit gas as determined from the stoichiometric reduction equation is equal to 67.258 kg moles

Difference 20.93 kg moles is due to excess O_2 and excessin N_2 the flue gas. Thus the revised composition of exit gas is

	Kg moles
Zn(v) =	6.173
O_2	4.40
CO_2 =	7.070
CO =	14.140
N_2 =	56.405

Liquid slag contains FeO = 2.778 kg moles and SiO_2 = 1.667 kg moles

$$\text{Amount of coke} = 254.52 \text{ kg}.$$

$$\text{Amount of air} \quad 71.40 \text{ kg moles}$$

Heat Balance

Coke and Zn concentrate enter at 1100K, while air at 298K. The exit gas discharges at 1300K and liquid slag and lead exits at 1600K

Thermo chemical data

$$-\Delta H^{\circ}_{Zno} = 83500 \text{ kcal / kg mol}$$

$$-\Delta H^{\circ}_{pbo} = 52500 \text{ kcal / kg mol}$$

Heat content in kcal / kg mole

$$H_{1100} - H_{298} \,|\, ZnO \quad 9500$$

$$H_{1100} - H_{298} \,|\, PbO \quad 10800$$

$$H_{1100} - H_{298} \,|\, FeO \quad 10280$$

$$H_{1100} - H_{298} \,|\, SiO_2 \quad 19940$$

$$H_{1100} - H_{298} \,|\, C \quad 3320$$

$$H_{1300} - H_{298} \,|\, Zn(v) \quad 36160$$

$$H_{1300} - H_{298} \,|\, CO_2 \quad 12010$$

$$H_{1300} - H_{298} \,|\, CO \quad 7460$$

$$H_{1300} - H_{298} \,|\, N_2 \quad 7500$$

$$H_{1300} - H_{298} \,|\, O_2 \quad 7873$$

$$H_{1600} - H_{298} \,|\, Pb \quad 10110$$

$$H_{1600} - H_{298} \,|\, FeO \quad 11990$$

$$H_{1600} - H_{298} \,|\, SiO_2 \quad 21100$$

Heat of formation of $2FeO.SiO_2 = 153 kcal/kg\,FeO$

One Can Perform Heat Balance

Heat input	(kcal)	Heat output	(kcal)
Heat of reaction	562539	Exit gas	871289
Sensible heat	200687	Lead	9069
Heat of slag formation	30602	Slag	68481
Total	793828	Heat losses*	79382 1028221

- Heat losses are normally taken as 10% of heat inputs if not mentioned in the problem.

We note there is heat deficit to the amount 234393 kcal. There are several ways to meet the heat deficit. One of the ways is to preheat the air. If average Cp of air is 7.5 kcal kg mole^{-1}°C^{-1} then Preheat air temperature is $463°C$.

Analysis of Heat Balance

54.8% of the total heat is supplied through the heat of reaction. Preheating of air and the reactants bring 22.8% and 19.5% respectively of the total heat requirements. Heat supplied due to formation of slag is 2.9% of total requirement.

Heat carried by Zn vapor = 21.7%; vapors are shocked and heat recovery is within lead bath.

Amount of heat is carried away by gases $CO + CO_2 + N_2$ which is equal to 63.1%. Once the gases are shock cooled, temperature of gases $CO + CO_2 + N_2$ must come down. Information on this temperature will help to consider waste heat recovery

Iron Making

Blast furnace iron making is an example in which both unit processes that is reduction of iron oxide and smelting to separate liquid pig iron from slag is done simultaneously in a single reactor. Height of the blast furnace is around 25 to 30m. It must be mentioned that blast furnace is a very efficient reactor both in terms of heat and mass exchange between solids and gases.

Coke (Fuel)

Raw coke

Coke is a fuel with few impurities and a high carbon content, usually made from coal. It is the solid carbonaceous material derived from destructive distillation of low-ash, low-sulfur bituminous coal. Cokes made from coal are grey, hard, and porous. While coke can be formed naturally, the commonly used form is man-made. The form known as petroleum coke, or pet coke, is derived from oil refinery coker units or other cracking processes.

Coke is used in preparation of producer gas which is a mixture of carbon monoxide (CO) and nitrogen (N_2). Producer gas is produced by passing air over red-hot coke. Coke is also used to manufacture water gas.

History

China

Historical sources dating to the 4th century describe the production of coke in ancient China. The Chinese first used coke for heating and cooking no later than the ninth century. By the first decades of the eleventh century, Chinese ironworkers in the Yellow River valley began to fuel their furnaces with coke, solving their fuel problem in that tree-sparse region.

Britain

In 1589 a patent was granted to Thomas Proctor and William Peterson for making iron and steel and melting lead with "earth-coal, sea-coal, turf, and peat". The patent contains a distinct allusion to the preparation of coal by "cooking". In 1590 a patent was granted to the Dean of York to "purify pit-coal and free it from its offensive smell". In 1620 a patent was granted to a company composed of William St. John, Robert Follensbee and other knights, mentioning the use of coke in smelting ores and manufacturing metals. In 1627 a patent was granted to Sir John Hacket and Octavius de Strada for a method of rendering sea-coal and pit-coal as useful as charcoal for burning in houses, without offense by smell or smoke.

In 1603 Hugh Plat suggested that coal might be charred in a manner analogous to the way charcoal is produced from wood. This process was not put into practice until 1642, when coke was used for roasting malt in Derbyshire; previously, brewers had used wood, as uncoked coal cannot be used in brewing because its sulfurous fumes would impart a foul taste to the beer. It was considered an improvement in quality, and brought about an "alteration which all England admired"—the coke process allowed for a lighter roast of the malt, leading to the creation of what by the end of the 17th century was called pale ale.

In 1709 Abraham Darby I established a coke-fired blast furnace to produce cast iron. Coke's superior crushing strength allowed blast furnaces to become taller and larger. The ensuing availability of inexpensive iron was one of the factors leading to the Industrial Revolution. Before this time, iron-making used large quantities of charcoal, produced by burning wood. As the coppicing of forests became unable to meet the demand, the substitution of coke for charcoal became common in Great Britain, and the coke was manufactured by burning coal in heaps on the ground in such a way that only the outer layer burned, leaving the interior of the pile in a carbonized state. In the late 18th century, brick beehive ovens were developed, which allowed more control over the burning process.

In 1768 John Wilkinson built a more practical oven for converting coal into coke. Wilkinson improved the process by building the coal heaps around a low central chimney built of loose bricks and with openings for the combustion gases to enter, resulting in a higher yield of better coke. With greater skill in the firing, covering and quenching of the heaps, yields were increased from about 33 per cent to 65 per cent by the middle of the 19th century. The Scottish iron industry expanded very rapidly in the second quarter of the 19th century, through the adoption of the hot-blast process in its coalfields.

In 1802 a battery of beehives was set up near Sheffield, to coke the Silkstone seam for use in crucible steel melting. By 1870, there were 14,000 beehive ovens in operation on the West Durham coalfields, capable of producing 4.2 million tons of coke. As a measure of the extent of the expansion of coke-making, it has been estimated that the requirements of the iron industry were about one million tons a year in the early 1850s, whereas by 1880 the figure had risen to seven millions, of which about 5 millions were produced in Durham county, one million tons in the South Wales coalfield, and 1 million tons in Yorkshire and Derbyshire.

In the first years of steam railway locomotives, coke was the normal fuel. This resulted from an early piece of environmental legislation; any proposed locomotive had to "consume its own smoke". This was not technically possible to achieve until the firebox arch came into use, but burning coke, with its low smoke emissions, was considered to meet the requirement. However, this rule was quietly dropped and cheaper coal became the normal fuel, as railways gained acceptance among the general public.

United States

Illustration of coal mining and coke burning from 1879

In the United States, the first use of coke in an iron furnace occurred around 1817 at Isaac Meason's Plumsock puddling furnace and rolling mill in Fayette County, Pennsylvania. In the late 19th century, the coalfields of western Pennsylvania provided a rich source of raw material for coking. In 1885, the Rochester and Pittsburgh Coal and Iron Company constructed the world's longest string of coke ovens in Walston, Pennsylvania, with 475 ovens over a length of 2 km (1.25 miles). Their output reached 22,000 tons per month. The Minersville Coke Ovens in Huntingdon County, Pennsylvania, were listed on the National Register of Historic Places in 1991.

Coal coking ovens at Cokedale, west of Trinidad, Colorado, supplied steel mills in Pueblo, Colorado

Between 1870 and 1905, the number of beehive ovens in the United States skyrocketed from about 200 to almost 31,000, which produced nearly 18 million tons of coke in the Pittsburgh area alone. One observer boasted that if loaded into a train, "the year's production would make up a train so long that the engine in front of it would go to San Francisco and come back to Connellsville before the caboose had gotten started out of the Connellsville yards!" The number of beehive ovens in Pittsburgh peaked in 1910 at almost 48,000.

Although it made a top-quality fuel, coking poisoned the surrounding landscape. After 1900, the serious environmental damage of beehive coking attracted national notice, even though the damage had plagued the district for decades. "The smoke and gas from some ovens destroy all vegetation around the small mining communities," noted W. J. Lauck of the U.S. Immigration Commission in 1911. Passing through the region on train, University of Wisconsin president Charles van Hise saw "long rows of beehive ovens from which flame is bursting and dense clouds of smoke issuing, making the sky dark. By night the scene is rendered indescribably vivid by these numerous burning pits. The beehive ovens make the entire region of coke manufacture one of dulled sky: cheerless and unhealthful."

Production

Volatile constituents of the coal—including water, coal-gas, and coal-tar—are driven off by baking in an airless furnace or oven (kiln) at temperatures as high as 2,000 °C (3,600 °F) but usually around 1,000–1,100 °C (1,800–2,000 °F). This fuses together the fixed carbon and residual ash. Some facilities have "by-product" coking ovens in which the volatile hydrocarbons are mainly used, after purification, in a separate combustion process to generate energy. Non by-product coking furnaces or coke furnaces (ovens) burn the hydrocarbon gases produced by the coke-making process to drive the carbonization process. This is an older method, but is still being used for new construction.

Coke oven at smokeless fuel plant, Abercwmboi, South Wales, 1976

Bituminous coal must meet a set of criteria for use as coking coal, determined by particular coal assay techniques. These include moisture content, ash content, sulfur content, volatile content, tar, and plasticity. This blending is targeted at producing a coke of appropriate strength (generally measured by Coke Strength After Reaction (CSR)), while losing an appropriate amount of mass. Other blending considerations include ensuring the coke doesn't swell too much during production and destroy the coke oven through excessive wall pressures.

The greater the volatile matter in coal, the more by-product can be produced. It is generally considered that levels of 26–29% of volatile matter in the coal blend are good for coking purposes. Thus different types of coal are proportionally blended to reach acceptable levels of volatility before the coking process begins.

Coking coal is different from thermal coal, but it differs not by the coal forming process. Coking coal has different macerals from thermal coal. The different macerals are related to source of material that compose the coal. However, the coke is of wildly varying strength and ash content and is generally considered unsellable except in some cases as a thermal product. As it has lost its volatile matter, it has lost the ability to be coked again.

The Hearth Process

The "Hearth" process of coke-making, using lump coal, was akin to that of charcoal-burning; instead of a heap of prepared wood, covered with twigs, leaves and earth, there was a heap of coals, covered with coke dust. The hearth process continued to be used in many areas during the first half of the 19th century, but two events greatly lessened its importance. These were the invention of the hot blast in iron-smelting and the introduction of the beehive coke oven. The use of a blast of hot air, instead of cold air, in the smelting furnace was first introduced by Neilson in Scotland in the year 1828. The hearth process of making coke from coal is a very lengthy process.

Beehive Coke Oven

Coke ovens and coal tipple in Pennsylvania

A fire brick chamber shaped like a dome is used, commonly known as a beehive oven. It is typically 4 meters wide and 2.5 meters high. The roof has a hole for charging the coal or other kindling from the top. The discharging hole is provided in the circumference of the lower part of the wall. In a coke oven battery, a number of ovens are built in a row with common walls between neighboring ovens. A battery consisted of a great many ovens, sometimes hundreds, in a row.

Coal is introduced from the top to produce an even layer of about 60 to 90 centimeters deep. Air is supplied initially to ignite the coal. Carbonization starts and produces volatile matter, which burns inside the partially closed side door. Carbonization proceeds from top to bottom and is completed in two to three days. Heat is supplied by the burning volatile matter so no by-products are recovered. The exhaust gases are allowed to escape to the atmosphere. The hot coke is quenched with water and discharged, manually through the side door. The walls and roof retain enough heat to initiate carbonization of the next charge.

When coal was burned in a coke oven, the impurities of the coal not already driven off as gases accumulated to form slag, which was effectively a conglomeration of the removed impurities. Since it was not the desired coke product, slag was initially nothing more than an unwanted by-product and was discarded. Later, however, it was found to have many beneficial uses and has since been used as an ingredient in brick-making, mixed cement, granule-covered shingles, and even as a fertilizer.

Occupational Safety

People can be exposed to coke oven emissions in the workplace by inhalation, skin contact, or eye contact. The Occupational Safety and Health Administration (OSHA) has set the legal limit for coke oven emissions exposure in the workplace as 0.150 mg/m$^{3\text{benzene}}$-soluble fraction over an 8-hour workday. The National Institute for Occupational Safety and Health (NIOSH) has set a Recommended exposure limit (REL) of 0.2 mg/m$^{3\text{benzene}}$-soluble fraction over an 8-hour workday.

Uses

Coke is used as a fuel and as a reducing agent in smeltingiron ore in a blast furnace. The carbon monoxide produced by its combustion reduces iron oxide (hematite) in the production of the iron product.

Coke is commonly used as fuel for blacksmithing.

Coke was used in Australia in the 1960s and early 1970s for house heating.

Since smoke-producing constituents are driven off during the coking of coal, coke forms a desirable fuel for stoves and furnaces in which conditions are not suitable for the complete burning of bituminous coal itself. Coke may be combusted producing little or no smoke, while bituminous coal would produce much smoke. Coke was widely used as a substitute for coal in domestic heating following the creation of smokeless zones in the United Kingdom.

Highland Park distillery in Orkney roasts malted barley for use in their Scotchwhisky in kilns burning a mixture of coke and peat.

Discovered by accident to have superior heat shielding properties when combined with other materials, coke was one of the materials used in the heat shielding on NASA's Apollo Command Module. In its final form, this material was called AVCOAT 5026-39. This material has been used most recently as the heat shielding on the Mars Pathfinder vehicle. Although it was not used for the space shuttle, NASA had been planning to use coke and other materials for the heat shield for its next generation space craft, Orion.

Coke may be used to make synthesis gas, a mixture of carbon monoxide and hydrogen.

- Syngas; water gas: a mixture of carbon monoxide and hydrogen, made by passing steam over red-hot coke (or any carbon-based char)

- Producer gas; suction gas; wood gas; generator gas; synthetic gas: a mixture of carbon monoxide, hydrogen, and nitrogen, made by passing air over red-hot coke (or any carbon-based char)

Phenolic Byproducts

Wastewater from coking is highly toxic and carcinogenic. It contains phenolic, aromatic, heterocyclic, and polycyclic organics, and inorganics including cyanides, sulfides, ammonium and ammonia. Various methods for its treatment have been studied in recent years. The white rot fungus *Phanerochaete chrysosporium* can remove up to 80% of phenols from coking waste water.

Properties

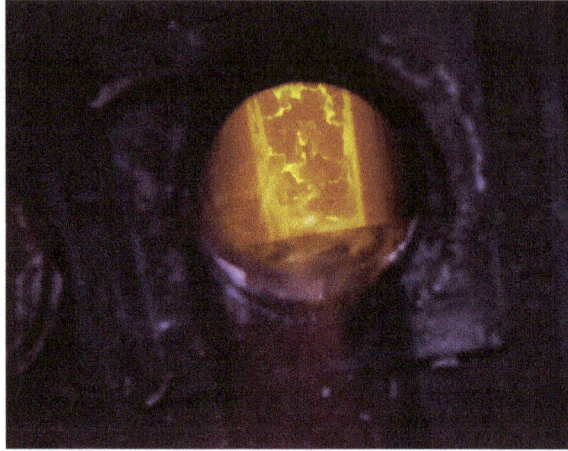

Hanna furnaces of the Great Lakes Steel Corporation, Detroit. Coal tower atop coke ovens.

The bulk specific gravity of coke is typically around 0.77. It is highly porous.

The most important properties of coke are ash and sulfur content, which are dependent on the coal used for production. Coke with less ash and sulfur content is highly priced on the market. Other important characteristics are the M10, M25, and M40 test crush indexes, which convey the strength of coke during transportation into the blast furnaces; depending on blast furnaces size, finely crushed coke pieces must not be allowed into the blast furnaces because they would impede the flow of gas through the charge of iron and coke. A related characteristic is the Coke Strength After Reaction (CSR) index; it represents coke's ability to withstand the violent conditions inside the blast furnace before turning into fine particles.

The water content in coke is practically zero at the end of the coking process, but it is often water quenched so that it can be transported to the blast furnaces. The porous structure of coke absorbs some water, usually 3–6% of its mass. In more modern coke plants an advanced method of coke cooling uses air quenching.

Bituminous coal must meet a set of criteria for use as coking coal, determined by particular coal assay techniques.

Other Processes

The solid residue remaining from refinement of petroleum by the "cracking" process is also a form of coke. Petroleum coke has many uses besides being a fuel, such as the manufacture of dry cells and of electrolytic and welding electrodes.

The Illawarra Coke Company (ICC) in Coalcliff, New South Wales, Australia.

Gas works manufacturing syngas also produce coke as an end product, called gas house coke.

Fluid coking is a process which converts heavy residual crude into lighter products such as naphtha, kerosene, heating oil, and hydrocarbon gases. The "fluid" term refers to the fact that coke particles are in a continuous system versus older batch-coking technology.

Materials Balance Calculation Procedure in Coke Making

In coke making coal of certain composition is carbonized in a by-product coke oven. As a result of carbonization, products and by-product are produced. Main product is coke, whereas by-products are coke oven gas and tar.

Basis of calculation: one may take 1 kg coal, 100kg coal or 1000kg coal.

Amount of coke is determined by ash balance, namely if W kg is amount of coke then ash balance is

$$A_1 \, 1000/100 = A_2 W/100 + A_3 \times \text{wt.of tar}/100$$

A_1 is the amount of ash in coal, A_2 and A_3 are amounts of ash in coke and tar respectively. By knowing weight of tar, W can be calculated.

Amount of coke oven gas can be calculated by carbon balance

$$\text{C from coal} = \text{C in coke} + \text{C in tar} + \text{C in coke oven gas}$$

Heat Balance

For heat balance calculations, reverence temperature of 298K is normally selected.

Heat balance at steady state is

$$\text{Heat input} = \text{Heat output} + \text{Heat loesses}$$

In coke –making heat input is the calorific value of coal and calorific value of coke- oven gas calorific value of coal and of any other solid fuel can be determined by Dulong formula:

$$GCV = 339\%C + 1427(\%H - \%O/8) + 93\%S \text{ KJ}/\text{kg}$$

$$NCV = GCV = 24.44(9\%H + \%M)kJ/kg$$

GCV is gross calorific value and NCV is the net calorific value.

Heat output consists of

1. Sensible heat in coke:

It can determined by $WC_{pc}(T_{coke} - 298)$ in Kcal or kj, where X is mass of coke, Cp is specific heat of coke and T_{coke} is temperature of coke discharged from coke oven. C_{pc} is 0.359 kcal / kg°C.

a) Sensible heat in coke oven gas: it may be calculated by $YC_{pg}(T_g - 298)$ in kcal or kj, where C_{pg} is specific heat of coke oven gas. Its value may be taken as 0.44 kcal / m³°C.

b) CV of coke and tar can be calculated by Dulong formula.

c) CV of coke-oven gas is the summation of heat of combustion values of all combustible components in coke-oven gas.

Heat balance calculations disclose

a) Distribution of heat energy in products and by-products.

b) Sensible heat available in products and by-products.

c) Heat losses can be determined from difference between heat output and heat input values.

Materials Balance In Iron Making

The blast furnace is essentially a continuous counter-current reactor in which the descending charge is heated and reacted with ascending gases, derived from combustion of carbon at the tuyere. The charge consists of iron sinter/pellets + coke and limestone. During descent, iron oxide is reduced to FeO and limestone decomposed to CaO and CO_2.

The combustion of coke at the tuyere level with air raises the temperature in between $1800°C$ to $2100°C$ and melts slag and metal.

The reduction of FeO to Fe and the melting of iron and formation of hot metal and slag begin once the charge descends to the bosh region. The following reduction reactions occur:

$$SiO_2 + 2C = Si + 2CO$$

$$MnO + C = Mn + CO$$

$$P_2O_5 + 5C = 2P + 5CO$$

$$FeO + C = Fe + CO$$

The gases consisting mainly of CO and N_2 and some H_2, derived from moisture of blast, ascends through the charge. Heat transfer occurs as the gas rising upward and simultaneously reduction of wustile to Fe occurs.The CO_2 so produced quickly reacts with C and produces CO $CO_2 + C = 2CO$

Further reaction between CO and higher oxides of iron will produce CO_2 which accumulates in the ascending gases. The content of CO_2 increases with the decomposition of $CaCO_3$ $CaCO_3 = CaO + CO_2$

at about $950°C$. The exit gas may contain CO/CO_2 ratio close to one and leaves the furnace at about 500-600K.

It may also be noted that the reduction of FeO to Fe occurs both by carbon (called direct reduction) and CO (called indirect reduction)

Example

Let me illustrate material balance through a problem

Consider a blast furnace which is charged with iron ore coke and flux of the following composition:

Iron ore (weight %):	Fe_2O_3 =78, SiO_2 -8.4, MnO = 0.6, Al_2O_3=5.0, P_2O_5 = 1.7MgO = 1.2 and H_2O=5.1
Coke (weight %):	C=88, SiO_2 =9, Al_2O_3 =1 and H_2O = 2
Flue :	$CaCO_3$=96 %, $MgCO_3$ =2% and SiO_2 =2%

Pig iron analyses (Weigh percent%) $= Fe = 92.7, C = 4Si = 2, P = 0.9$ and $Mn = 0.4$

The coke rate is 900kg/ton of pig iron. (Modern blast furnace operates with much lower coke rate)

During smelting 99.5% of Fe is reduced and 0.5% is slagged. The CO/CO_2 ratio in the top gas is 2/1

Calculate

1.Weight of iron ore

2.Weight and composition of slag

3.Volume of air required

4.Volume and% composition of exit gas.

Solution

Fe balance: $(0.995).(\text{Fe in iron ore}) = \text{Fe in pig iron}$

Let x kg iron ore

$$0.995 \times \frac{112}{160} \times 0.78 \times = 927$$

$x = 1706 \text{ kg Ans (a)}$

Weight of Slag

Slag consists of $FeO, SiO_2, Al_2O_3, MgO, P_2O_5, CaO$

Si in slag = Si in ore + Si in coke + Si in limestone = Si in pig iron

$$= 88.65\text{kg} = 3.166 \text{ kg moes}$$

$$SiO_2 \text{ in slag} = 189.97 \text{ kg}$$

$$Al_2O_3 \text{ in slag} = 0.05 \times 1706 + 900 \times 0.01$$

$$= 94.3 \text{ kg}$$

$$MnO \text{ in slag} = \frac{71}{55}[\text{Mn from ore} - \text{Mn in pig iron}]$$

$$= 5.04\text{kg}$$

$$P_2O_5 \text{in slag} = \frac{142}{62}[\text{P from ore} - \text{P in pig iron}]$$

$$= 8.4\text{kg}$$

$$MgO \text{ in slag} = 24.53 \text{ kg}$$

$$CaO \text{ in slag} = 229\text{kg}$$

$$FeO \text{ in slag} = 6\text{kg}$$

Weight of Slag and its Composition in Percent in Given Below

	Kg	%
SiO$_2$	189.97	34.09
FeO	6.00	1.08
Al$_2$O$_3$	94.30	16.92
MnO	5.04	0.92
P$_2$O$_5$	8.40	1.50
MgO	24.53	4.41
CaO	229.00	41.10
Total	557.24	100%

Volume of air required calculation to consider the top gas. All the carbon charged except that dissolves in iron will be available in the top gas. Air is required in the blast furnace to combust carbon of coke at the tuyere level. From the amount of coke and the oxygen available through the reduction of oxides, volume of air can be determined.

Oxygen available from the charge = Decomposition of $CaCO_3$ and $MgCO_3$ + oxygen released through the reduction of the oxides

Total O_2 available from charge $= 17.742$ kg moles

$$\text{Total C in gases} = \frac{0.88 \times 900}{12} + \left[\frac{0.96}{100} + \frac{0.02}{84}\right]\frac{1706}{4.} - \frac{40}{12}$$

$$= 66.863 \text{ kg moles}$$

C in to $CO = 44.575$ kg moles O_2 required $= 22.2875$

C to $CO_2 = 22.288$ kg moles O_2 required $= 22.288$

Total $= 44.5755$ kg moles

O_2 from air $= 44.5755 - 17.742 = 26.8335$

Air $= 2862.24\text{m}^3$

Volume and % composition of flue gas

	Kgmoles	%
Co =	44.575	25.82
CO_2 =	22.288	12.91
N_2 =	100.945	58.47
H_2O =	4.834	2.80

Total $172.642\ kg\ moles$ 100%

Also important is the amount of carbon burnt at the tuyere level.

In order to find amount of carbon, burnt at the tuyere, volume of blast can be used.

N_2 of blast is inert and O_2 react with carbon only.

All oxygen of blast reacts with C and forms CO

Volume of blast $= 2862.24\text{m}^3 = 127.8$ kg moles

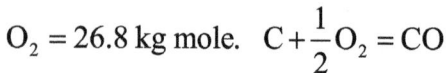

$O_2 = 26.8$ kg mole. $C + \dfrac{1}{2}O_2 = CO$

$C = 53.7$ kg mole $= 644$ kg

%C. burnt at tuyere level $= \dfrac{644}{9000 \times 0.88} = 81.3\%$

References

- Lu, Y; Yan, L; Wang, Y; Zhou, S; Fu, J; Zhang, J (2009). "Biodegradation of phenolic compounds from coking wastewater by immobilized white rot fungus Phanerochaete chrysosporium". Journal of Hazardous Materials. 165 (1–3): 1091–97. PMID 19062164. doi:10.1016/j.jhazmat.2008.10.091

- Nersesian, Roy L (2010). "Coal and the Industrial Revolution". Energy for the 21st century (2 ed.). Armonk, NY: Sharpe. p. 98. ISBN 978-0-7656-2413-0

- "CCHC—Your Portal to the Past". Coal and Coke Heritage Center. Penn State Fayette, The Eberly Campus. Retrieved 19 March 2013

- Wittcoff, M.M. Green ; H.A. (2003). Organic chemistry principles and industrial practice (1. ed., 1. reprint. ed.). Weinheim: Wiley-VCH. ISBN 3-527-30289-1

- "201006274431 | Belovode site in Serbia may have hosted first copper makers". Retrieved 26 August 2015

Furnace Operations in Metallurgy

The chapter talks about furnaces such as blast furnace and cupola furnace and their related stoichiometries. A blast furnace is a high-temperature heating device that is used in the process of smelting. A cupola furnace is used to melt cast iron, and unlike a blast furnace, it could vary in size.

Blast Furnace Stoichiometry

In blast furnace iron oxide +coke + flux are charged from top and air is injected through the tuyeres located near the hearth region.

The gases CO, CO_2, and N_2 leave the furnace through the top. Some amount of carbon in dissolved in iron. Pig iron and slag leave the bottom.

Under simplified conditions, one can derive equation to represent stoichiometry in blast furnace on a diagram. Such diagram is called RIST diagram.

Stoichiometry in blast furnace:

Following assumptions are made:

i) iron of iron ore enters in hot metal (or pig iron)

ii) Oxygen enters through air blast and oxides of iron.

iii) Carbon enters through coke.

Consider 1 mole of iron

$$n^i_{Fe} = n^o_{Fe} = 1$$

$$n^i_c = n^o_c$$

n^i_{Fe}, n^o_{Fe} = Number of moles of iron entering and leaving the furnace

n^o_c = Number of moles of carbon in gas phase + number of moles of carbon in Fe

$$n^o_c = (n_c)_g + \left(\frac{c}{Fe}\right)^m$$

By above given two equations we get,

$$n_c^i = (n_c)_g + \left(\frac{c}{Fe}\right)^m$$

Similarly oxygen balance

$$n_o^i = n_o^o$$

n_o^i = Number of moles of oxygen entering

$$= (n_o)_B + \left(\frac{o}{Fe}\right)^x$$

Where $(n_o)_B$ moles of O_2 entering through air and $\left(\frac{O}{Fe}\right)^x$ moles of oxygen with Fe_2O_3 or Fe_3O_4 as the case may be

Since all oxygen leaves as CO or CO

$$n_o^i = n_o^o = (n_c)_g \times \left(\frac{o}{c}\right)_g$$

By above given two equations we get,

$$(n_o)_B + \left(\frac{o}{Fe}\right)^x = (n_c)_g \times \left(\frac{o}{c}\right)_g$$

$\left(\frac{o}{Fe}\right)^x$ and $\left(\frac{o}{c}\right)$ depends on incoming iron oxide and outgoing gases, for example

$$\left(\frac{o}{Fe}\right)^x = 1.5 \text{ in } Fe_2O_3$$

Of the total carbon charged in the blast furnace a fraction reacts with oxygen and this carbon is called active carbon (n_A^C)and the rest fraction dissolves in iron and this carbon is termed as inactive carbon; *i.e.* $\left(\frac{C}{Fe}\right)^m$. In equation 8 ($n_g^c$) corresponds to the carbon which has reacted with oxygen and exited the furnace; and hence equal to (n_A^C). With this argument equation 8 modifies to

$$(n_o)_B + \left(\frac{o}{Fe}\right)^x = n_c^A \left(\frac{o}{c}\right)_g$$

Note that $n_c^A = n_i^c - \left(\frac{c}{Fe}\right)^m$

Example

Calculate top gas composition for an ideal gas blast furnace operating with a mixture of Fe_2O_3 and coke. Coke is 90%C and is consumed at the rate 475kg | ton of iron. Oxygen is introduced at the rate of 350kg / 1000kg iron. Hot metal has 4.5%C.

Blast in traduces oxygen at 350kg / 1000kg iron. Hot metal has 4.5%C.

Basis of calculation $= 1000$ kg iron or 17.9 kg moles.

In Fe_2O_3 $\left(\dfrac{O}{Fe}\right)^X = 1.5$

Note that hot metal has 95.5% Fe and 4.5%C

In hot metal $C = 47$ kg $\therefore \left(\dfrac{C}{Fe}\right)^m = 0.219$

In Coke: carbon 428 kg, and hence $n_c^i = 35.6$ kg moles / ton Fe

Active carbon $n_c^A = 1.77$

Oxygen from blast $n_o^B = 1.22$ kg atams / moles of Fe

By using equation $(n_o)_B + \left(\dfrac{O}{Fe}\right)^X = n_c^A \left(\dfrac{O}{c}\right)g$ we get

$$\left(\frac{O}{C}\right)g = 1.54$$

$$\times_{CO_2}^g = \left(\frac{O}{C}\right)^g - 1$$

$$\times_{CO}^g = 2 - \left(\frac{O}{C}\right)^g$$

$\times_{CO_2}^g$ and $\times_{CO}^g$ are mole fraction of CO and CO.

Moles of $n_{CO_2}^g = n_c^A \times (\times_{CO_2}^g) = 0.96$ moles / mole of Fe

$$n_{CO}^g = n_c^A \times (\times_{CO}^g) = 0.81 \text{mole / mole of Fe}$$

$$N_2 = \frac{0.79}{0.21} \times \frac{1}{2} \times 1.22 = 2.29 \text{mole / mole of Fe}$$

Top gas composition: 20.0 vol%CO

$$23.6 \text{ vol}\%CO_2$$

$$56.4 \text{ vol}\%N_2$$

Graphical Representation of Stoichiometric Balance Equation

Writing Above Equation as:

$$\left(\frac{O}{Fe}\right)^{X} - (-n_B^O) = n_C^A \left\{\left(\frac{O}{C}\right)g^{-O}\right\}$$

$\dfrac{O}{Fe}$ and n_B^O are atoms of oxygen /mole of iron.

Equation 10 is equivalent to

$$Y_2 - Y_1 = M\{(X_2 - X_1)\}$$

The above equation is straight line of slope M passing through $x_1 Y_1$ and $x_2 Y_2$

Thus we can plot $(O|Fe)^x$ vs $(O/C)_g$ which give a straight line with slope (n_c^A). This plot is shown in the figure below.

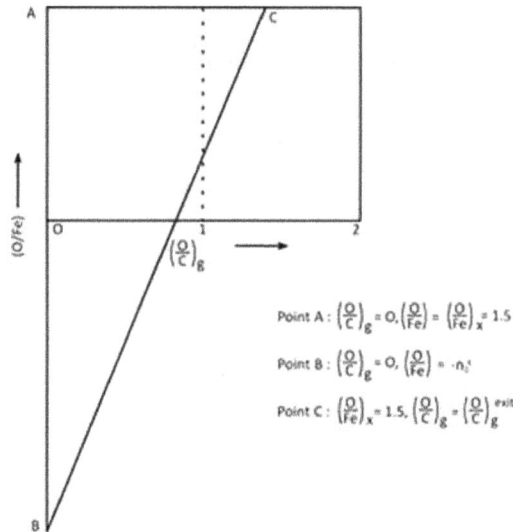

Point A: $\left(\frac{O}{C}\right)_g = 0, \left(\frac{O}{Fe}\right) = \left(\frac{O}{Fe}\right)_x = 1.5$

Point B: $\left(\frac{O}{C}\right)_g = 0, \left(\frac{O}{Fe}\right) = -n_c^o$

Point C: $\left(\frac{O}{Fe}\right)_x = 1.5, \left(\frac{O}{C}\right)_g = \left(\frac{O}{C}\right)_g^{exit}$

Plot of $(O/Fe)^x$ Vs $(O/C)_g$. The slope of the line is n_c^A.

In the figure $(O/Fe)^x$ is equal to 1.5 for Fe_2O_3. This point is fixed on the ordinate as shown by A. point B on the ordinate is $-n_B^O$ according to equation 10. X-axis is $(O/C)_g$.

Exit gas composition varies in between CO and CO_2, that is $\left(\dfrac{O}{C}\right)_g - 1$ and 2.

Point C is $(O/C)_g$ at $(O/Fe)^x = 1.5$

Example

The exit gas composition from a Fe_2O_3 charged furnace is $24vol\%CO, 22vol\%CO_2, 54\%N_2$. The air blast is $1400m^3/100kg$ of product Fe. The hot metal contains $5\%C$. Calculate:

a) Quantity of active carbon in kg | ton Fe

b) Total carbon in kg

Solution

We know

$$n_O^B + \left(\frac{O}{Fe}\right)^x = n_C^A \left(\frac{O}{C}\right)_g$$

Oxygen from blast $= 13.125$ kg moles .

$\therefore n_O^B = 1.466$ oxygen atom /mole Fe.

$$\frac{\times co}{\times co_2} = \frac{2 - \left(\frac{O}{C}\right)_g}{\left(\frac{O}{C}\right)_g - 1}$$

We get $\left(\frac{O}{C}\right)_g = 1.476$.

Substituting the value of $n_O^B, \left(\frac{O}{Fe}\right)^x$ and $\left(\frac{O}{C}\right)_g$ in equation 1 we get

$$n_C^A = 431 \text{ kg}$$

$$nc_i = n_C^A + \left(\frac{C}{Fe}\right)_m$$

C in kg $= 484$

Alternatively one can plot) (O / Fe) against $\left(\frac{O}{C}\right)_g$.

$$\left(\frac{O}{Fe}\right)^x = 1.5 \text{ and } \left(\frac{O}{C}\right)_g = 0. \left(\frac{O}{Fe}\right)^x = -n_O^B.$$

At $\left(\frac{O}{Fe}\right)^Y = 1.5, \left(\frac{O}{C}\right)_g = 1.476$

The straight line gaining the values of $-n_O^B$ and $(O / C)_g$ is the operating line of the furnace. Its slope is n_C^A Plot yourself as an exercise.

A simplified material balance is considered for ironmaking blast furnace and the results of material balance formulation are presented in the form RIST diagram. In that it is considered that charge consists of pure iron oxide and coke and the product is hot metal containing carbon only. We have not considered heat balance for that simple case.

This chapter considers heat balance in terms of heat input and output. Heat input is required to meet the thermal demand of the process of ironmaking. Heat demand of the blast furnace is met principally by the combustion of coke. In the calculations of coke rate, it is therefore necessary to consider the heat requirements. In the following we will be considering a simplified case in which pure oxide is a source of iron supply and coke is the source of carbon to illustrate the concept. This means that slag formation in this simplified case does not occur.

Enthalpy Balance in Blast Furnace

Consider a case where input is pure Fe_2O_3 (in reality iron ore is charged)and carbon (actually coke is used) at $298\,K$. Air is supplied at $298\,K$ (in reality air is preheated). Hot metal is considered to be a molten mixture of iron and carbon and exits at $1800\,K$. At the moment we exclude slag formation. Top gas leaves at $298\,K$ (this is an ideal case to explain the concept).

Enthalpy Balance

Enthalpy into furnace per mole of product Fe = Enthalpy out of furnace per mole of product Fe .

$$n_{Fe_2}O_3 \times \underset{Fe_2O_3}{H^\circ_{298}}, = \underset{Fe(1)}{H^\circ_{1800}} + n^g_{CO}.H^\circ_{298} + n^g_{CO2}H^\circ_{298}$$

Where

$n_{Fe_2}O_3, n^g_{CO}$ and $n^g_{CO_2}$ are moles of iron oxide, CO and CO_2 respectively | moles of products iron or

Or

$$\underbrace{n_{Fe_2}O_3 \times \left(\underset{Fe_2O_3}{-H^\circ_{298}} \right) + \underset{FeI.}{H^\circ_{1800}}}_{\text{Heat demand}} = \underbrace{n^g_{CO}(-H^\circ_{298}) + n^g_{CO}(-H^\circ_{298})}_{\text{Heat supply}}$$

Negative sign in above equation indicates that heat is produced. In the above equation heat is required to raise the temperature of hot metal from 298K to 1800K. Heat of formation of Fe_2O_3, CO and CO_2 is given below:

$$H^{\circ}_{298}(Fe_2O_3) = H^{f}_{298} = -826000 \text{ kJ} \,|\, kg \text{ mole} Fe_2O_3$$

$$H^{\circ}_{298}(CO_2) = H^{f}_{298} = -394000 \text{ kJ} \,|\, mole \text{ } CO_2$$

$$H^{\circ}_{298}(CO) = H^{f}_{298} = -111000 \text{ kJ} \,|\, kg \text{ mole } CO$$

$$H^{\circ}_{1800}(Fe_1) = \left[\underset{Fe(1)}{H^{\circ}_{1800}} - H^{\circ}_{298}Fe(s) \right] = -73000 \text{ kJ} \,|\, kg \text{ mole Fe} \cdot$$

Substituting the values in $\underbrace{n_{Fe_2O_3} \times \left(\underset{Fe_2O_3}{-H^{\circ}_{298}} \right) + \underset{Fe1.}{H^{\circ}_{1800}}}_{\text{Heat demand}} = \underbrace{n^{g}_{CO}(-H^{\circ}_{298}) + n^{g}_{CO}(-H^{\circ}_{298})}_{\text{Heat supply}}$ equation we get

$$n_{Fe_2O_3} \times 826000 + 73000 = n^{g}_{CO} \, 111000 + n^{g}_{CO_2} \, 394000$$

It is shown that:

$$n^{g}_{CO} = n^{g}_{C} X^{g}_{CO} = n^{A}_{c} \left(2 - (\%_C)g \right)$$

$$n^{g}_{CO_2} = n^{g}_{C} X^{g}_{CO_2} = n^{A}_{C} \left(\left(\tfrac{O}{C}\right)_g - 1 \right)$$

$$n_{Fe_zO_3} = \frac{1}{2} \text{(one mole of Fe requires } \tfrac{1}{2} \text{ mole of } Fe_2O_3)$$

By above given equations and noting $n_{Fe_2O_3} = \tfrac{1}{2}$

$$\frac{1}{2}.826000 + 73000 = n^{A}_{c} \left\{ 283000 \left(\frac{O}{c} \right)^{g} - 172000 \right\}$$

The above equation is in terms of kj /kg mole of product iron. Above equation is a simplified approach. It considers heat demand for hot metal only. In reality one has to consider heat demand for slag formation, decomposition of limestone, heat of mixing etc. The heat demand terms can be added simply on the left hand side of the above equation.

Left hand side of above equation is heat demand (D) which is a variable; variable because it may vary from furnace to furnace type of charge etc. Thus above equation is

$$D = n^{A}_{c} \left\{ 283000 \left(\frac{O}{c} \right)^{g} - 172000 \right\}.$$

D= heat demand of the process /kg mole of iron produced.

Illustration of the Concept

The CO/CO_2 in top gas leaving a hematite changed BF is ≈ 1. The carbon supply rate (incluing C in PI 5Wt.%) is 500 kg /ton of product iron. What is the enthalpy supply to the furnace (kj/kg mole of pnduct Fe) assuming blast and top gas leave at 298 K

Enthalpy supply $n_c^A \left\{ 283000 \left(\dfrac{O}{c} \right)^g - 172000 \right\}$.

$$\frac{Xco}{Xco_2} = \frac{2 - \left(\frac{O}{C} \right)_g}{\left(\frac{O}{C} \right)_g - 1} = 1$$

$$\left(\tfrac{O}{C} \right)_g = 1.5$$

$$n_C^i = 2.33 \text{ and } \left(\frac{C}{Fe} \right)_m = 0.025$$

$$n_C^A = 2.33 - 0.25 = 2.08$$

Enthalpy supply = 525200kj/g mole.

Earlier equation considers pure Fe as liquid product. On the demand side we can add more terms depending on the conditions.

a) Calculate heat demand kj /kg mole Fe when pure c and pure Fe_2O_3 enter at 298 K and liquid Fe exits at 1800k. D = 486000 kJ

b) Let iron contains 5%C . Neglecting heat of mixing

$$\left(\frac{C}{Fe} \right)_m = 0.25 \text{kg mol / mole of product iron}.$$

Enthalpy of C at 1800k = 7625 kJ

Enthalpy of Fe 5%C = 486000 + 7625 $\approx$ 49400 kj.

c) Let us consider that Fe 5% C contain 1%Mn and 1%Si

$$\left(\frac{Mn}{Fe} \right)^m = 0.011 \text{ kg mole Mn / kg mole Fe}$$

$$\left(\frac{Si}{Fe} \right)^m = 0.021 \text{ kg mole Si | kg mole Fe}$$

Heat demand in addition to $Fe\ 5\%C$ will consist of enthalpy of Mn, enthalpy of Si and heat required to reduce SiO_2 and MnO_2 to Si and Mn respectively.

Heat demand $= 521305\ kj/kg$ mole of iron.

Coupling of material and heat balance

Material balance equation is as follows:

$$n_O^B + \left(\frac{O}{Fe}\right)^x = n_C^A \left(\frac{O}{C}\right)_g$$

$$D = S = n_C^A[283000(\tfrac{O}{C})^g - 172000]$$

By above equations,

$$n_O^B + \left(\frac{O}{Fe}\right)^x - \frac{D}{283000} = n_C^A \frac{172000}{283000}$$

If $\left(\frac{O}{Fe}\right)$ and D are specified, then specification of either n_C^A or n_O^B fully defines blast furnace operation

Example

Consider pure Fe_2O_3 as a feed, hot metal exits at 1800k and hot metal contains 5%C. The value of $n_B^O = 1.41$ Substituting the values of $n_B^O, (\tfrac{O}{Fe})^x = 1.5$ and $D = 494000$.

$$n_C^A = 1.91$$

$$n_i^C = \left(\tfrac{C}{Fe}\right)^m + n_C^A = 2.16.$$

Carbon consumption $= 464kg\,|\,1000kg$ product Fe.

Graphical representation

Let us rewrite equation 9 as

$$\left\{ \left(\frac{O}{Fe}\right)^x - \frac{D}{283000} \right\} - (-n_O^B) = n_C^A \left\{ \frac{172000}{283000} - O \right\}$$

$$Y_2 - Y_1 = M(X_2 - X_1).$$

One can plot $\left(\frac{O}{Fe}\right)$ as ordinate against $(O/C)g$. The slope of the line n_C^A will pass through points

$$\left(\frac{O}{C}\right) = 0, \left(\frac{O}{Fe}\right) = -n_B^O$$

$$\left(\frac{O}{C}\right) = 0.61 \frac{O}{Fe} = \left(\frac{O}{Fe}\right)^x - \frac{D}{283000} = -0.57$$

Example 2

A blast furnace operates with 600kg C and pig iron Contains 5%C . feedconsists of Fe_2O_3 + gangue.

$$n_C^A = 2.55$$

We draw a line with slope 2.55, we determine

$$\left(\frac{O}{C}\right)_g = 1.42$$

Top gas composition $CO = 58\%, CO_2 = 42\%$.

Converting (Metallurgy)

Converting is a type of metallurgical smelting that includes several processes; the most commercially important form is the treatment of molten metal sulfides to produce crude metal and slag, as in the case of copper and nickel converting. A now-uncommon form is batch treatment of pig iron to produce steel by the Bessemer process. The vessel used was called the Bessemer converter.

Converting in Copper Metallurgy

A mixture of copper and iron sulfides referred to as matte is treated in converters to oxidize iron in the first stage, and oxidize copper in the second stage. In the first stage oxygen enriched air is blown through the tuyeres to partially convert metal sulfides to oxides:

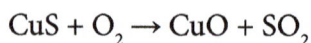

$$FeS + O_2 \rightarrow FeO + SO_2$$

$$CuS + O_2 \rightarrow CuO + SO_2$$

Since iron has greater affinity to oxygen, the produced copper oxide reacts with the remaining iron sulfide:

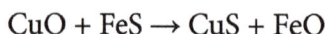

$$CuO + FeS \rightarrow CuS + FeO$$

The bulk of the copper oxide is turned back into the form of sulfide. In order to separate the obtained iron oxide, flux (mainly silica) is added into the converter. Silica reacts with iron oxide to produce a light slag phase, which is poured off through the hood when the converter is tilted around the rotation axis:

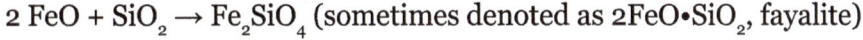

$$2\ FeO + SiO_2 \rightarrow Fe_2SiO_4\ (\text{sometimes denoted as } 2FeO \cdot SiO_2, \text{ fayalite})$$

After the first portion of slag is poured off the converter, a new portion of matte is added, and the converting operation is repeated many times until the converter is filled with the purified copper sulfide. The converter slag is usually recycled to the smelting stage due to the high content of copper in this by-product. Converter gas contains more than 10% of sulfur dioxide, which is usually captured for the production of sulfuric acid.

The second stage of converting is aimed at oxidizing the copper sulfide phase (purified in the first stage), and produces *blister copper*. The following reaction takes place in the converter:

$$CuS + O_2 \rightarrow Cu + SO_2$$

Copper content in the obtained blister copper is typically more than 95%. Blister copper is the final product of converting.

Equipment

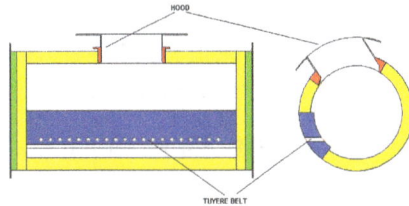

Sketch of a Pierce-Smith converter

The converting process occurs in a *converter*. Two kinds of converters are widely used: horizontal and vertical.

Horizontal converters of the Peirce-Smith type (which are an improvement of the Manhès-David converter) prevail in the metallurgy of non ferrous metals. Such a converter is a horizontal barrel lined with refractory material inside. A hood for the purpose of the loading/unloading operations is located on the upper side of the converter. Two belts of tuyeres come along the axis on either sides of the converter.

Molten sulfide material, referred to as matte, is poured through the hood into the converter during the operation of loading. Air is distributed to tuyeres from the two tuyere collectors which are located on opposite sides of the converter. Collector pipes vary in diameter with distance from the connection to air supplying trunk; this is to provide equal pressure of air in each tuyere.

Cupola Furnace

A small cupola furnace in operation at Wayne State University, in Detroit, Michigan.

A cupola or cupola furnace is a melting device used in foundries that can be used to melt cast iron, Ni-resist iron and some bronzes. The cupola can be made almost any practical size. The size of a cupola is expressed in diameters and can range from 1.5 to 13 feet (0.5 to 4.0 m). The overall shape is cylindrical and the equipment is arranged vertically, usually supported by four legs. The overall look is similar to a large smokestack.

The bottom of the cylinder is fitted with doors which swing down and out to 'drop bottom'. The top where gases escape can be open or fitted with a cap to prevent rain from entering the cupola. To control emissions a cupola may be fitted with a cap that is designed to pull the gases into a device to cool the gases and remove particulate matter.

The shell of the cupola, being usually made of steel, has refractory brick and plastic refractory patching material lining it. The bottom is lined in a similar manner but often a clay and sand mixture ("bod") may be used, as this lining is temporary. Finely divided coal ("sea coal") can be mixed with the clay lining so when heated the coal decomposes and the bod becomes slightly friable, easing the opening up of the tap holes. The bottom lining is compressed or 'rammed' against the bottom doors. Some cupolas are fitted with cooling jackets to keep the sides cool and with oxygen injection to make the coke fire burn hotter.

History

Cupola furnaces were built in China as early as the Warring States period (403–221 BC), although Donald Wagner writes that some iron ore melted in the blast furnace may have been cast directly into molds. During the Han Dynasty (202 BC – 220 AD), most, if not all, iron smelted in the blast furnace was remelted in a cupola furnace; it was designed so that a cold blast injected at the bottom traveled through tuyere pipes across the top where the charge (i.e. of charcoal and scrap or pig iron) was dumped, the air becoming a hot blast before reaching the bottom of the furnace where the iron was melted and then drained into appropriate molds for casting. A cupola furnace was made by René-Antoine Ferchault de Réaumur around 1720.

Fig. 68. – Foundry Cupola with Drop Bottom.

Drop-bottom cupola furnace

Operation

To begin a production run, called a 'cupola campaign', the furnace is filled with layers of coke and ignited with torches. Some smaller cupolas may be ignited with wood to start the coke burning. When the coke is ignited, air is introduced to the coke bed through ports in the sides called tuyeres.

When the coke is very hot, solid pieces of metal are charged into the furnace through an opening in the top. The metal is alternated with additional layers of fresh coke. Limestone is added to act as a flux. As the heat rises within the stack the metal is melted. It drips down through the coke bed to collect in a pool at the bottom, just above the bottom doors. During the melting process a thermodynamic reaction takes place between the fuel and the blast air. The carbon in the coke combines with the oxygen in the air to form carbon monoxide. The carbon monoxide further burns to form carbon dioxide. Some of the carbon is picked up by the falling droplets of molten metal which raises the carbon content of the iron. Silicon carbide and ferromanganese briquettes may also be added to the charge materials. The silicon carbide dissociates and carbon and silicon enters into the molten metal. Likewise, the ferromanganese melts and is combined into the pool of liquid iron in the 'well' at the bottom of the cupola. Additions to the molten iron such as ferromanganese, ferrosilicon, Silicon carbide and other alloying agents are used to alter the molten iron to conform to the needs of the castings at hand.

The operator of the cupola is known as the "cupola tender" or "furnace master". During the operation of a tapped cupola (cupolas may vary in this regard) the tender observes the amount of iron rising in the well of the cupola. When the metal level is sufficiently high, the cupola tender opens the "tap hole" to let the metal flow into a ladle or other container to hold the molten metal. When enough metal is drawn off the "tap hole" is plugged with a refractory plug made of clay.

The cupola tender observes the furnace through the sight glass or peep sight in the tuyeres. Slag will rise to the top of the pool of iron being formed. A slag hole, located higher up on the cylinder of the furnace, and usually to the rear or side of the tap hole, is opened to let the slag flow out. The viscosity is low (with proper fluxing) and the red hot molten slag will flow easily. Sometimes the slag which runs out the slag hole is collected in a small cup shaped tool, allowed to cool and harden. It is fractured and visually examined. With acid refractory lined cupolas a greenish colored slag means the fluxing is proper and adequate. In basic refractory lined cupolas the slag is brown.

After the cupola has produced enough metal to supply the foundry with its needs, the bottom is opened, or 'dropped' and the remaining materials fall to the floor between the legs. This material is allowed to cool and subsequently removed. The cupola can be used over and over. A 'campaign' may last a few hours, a day, weeks or even months.

When the operation is over, the blast is shut off and the prop under the bottom door is knocked down so that the bottom plates swing open. This enables the cupola remains to drop to the floor or into a bucket. They are then quenched and removed from underneath the cupola.

Quality Control

During the production, samples may be taken from the metal and poured into small molds. A chill wedge is often poured to monitor the iron quality. These small, approx 18 mm (3/4") wide x 38 mm (1-1/2") tall triangular shaped pieces are allowed to cool until the metal has solidified. They are then extracted from the sand mold and quenched in water, wide end first. After cooling in this manner the wedges are fractured and the metal coloration is assessed. A typical fracture will have a whitish color towards the thin area of the wedge and grayish color towards the wide end. The width of the wedge at the point of demarcation between the white and gray areas is measured and compared to normal results for particular iron tensile strengths. This visual method serves as a control measurement.

Operation of a Cupola

Cupola is used to melt pig iron to make iron castings in foundry. Charge consists of pig iron, scrap and limestone and coke. The oxidation of the following elements takes place:

$$Si + O_2 = SiO_2$$

$$Mn + 0.5\,O_2 = MnO$$

$$Fe + 0.5\,O_2 = FeO$$

SiO_2, MnO and FeO are slagged. Loss of carbon from pig iron may occur due to oxidation of carbon. However, this loss of carbon is compensated by absorption of carbon from coke. Cupola runs intermittently.

Air is blown through the tuyeres.

In material balance, the input consists of pig iron, scrap, limestone, coke and air. Whereas output consists of cast iron, slag and exit gases CO, CO_2, N_2 etc.

In cupola melting the calculation on material balance is required to determine the amounts of various inputs and outputs such that material management can be done for smooth inflow and output of materials.

Charge Balance Calculations

A cupola melts per hour 15 taons of pig iron of composition C 3.5%, Si 2.2% Mn 0.8% PO.7% and rest Fe; and 5 tons of scrap containing C 3,%, Si.8%MN 1.1% and PO.2%.

The dry air used is $849.6m^3$ measured at $313K$ to melt 1ton of pig iron and scrap per minute.

During melting 20% of total Si charged, 15% of total charged 1% of total Fe charged and 5% of C is oxidized, 19% of carbon of coke is absorbed by iron during melting. Enough $CaCO_3$ is charged to give 30% CaO in slag. The coke is 92% C and 8%SiO_2 and weight of coke is $1/9$ of the total weight of pig iron and scrap.

Required

a) Charge balances of cupola for 5 hr run.

b) The % composition of resulting cast iron, slag and gases.

Solution

Pig iron $= 250$ kg | min. and scrap $= 83.3$kg | min.

Metallic charge $= 333.3$kg | min.

Air blast $= 246.98m^3$ | min. out of which amount of $O_2 = 2.32$kg moles

The calculation on material balance is given below

Element	Amount charged(kg)	Amount oxidized (kg)
Si	7.0	1.4

Mn	2.92	0.44
Fe	310.20	3.10
C	9.25	0.46
P	1.92	-

From the amount of element oxidized one can calculate weight of slag

$= 15 \text{ kg} | \text{min}.$ (slag consists of $SiO_2 + MnO + FeO + CaO$)

Let us per form carbon and oxygen balance to calculate exit gas.

C from coke + C oxidized from pig iron + C from scrap + C in Ca CO_3 = C in gases

Oxygen from blast + oxygen from Ca CO_3 = oxygen in gases

Let x kg mole CO and Y kg mole CO_2 in exit gas

Equation

$$x + Y = 2.9234$$

$$x + 2Y = 4.72$$

we get $x = 1.1268$ and $Y = 1.7966$

Exit gas amount $= 1.1268 + 1.7966 + 8.7232 = 11.6466$ kg mole

% $CO = 9.67$, % $CO_2 = 15.43$ and % $N_2 = 74.90$

Exit gas volume $= 260.88 \text{ m}^3 | \text{minute}$

Amount of cast iron $= 329.29 \text{ kg} | \text{min}$

Charge Balance for 5 hr Operation

Pig iron $= 75$ ton   5 lag $= 4.5$ tons

Scrap $= 25$ ton   cast iron $= 98.8$ tons

Blast $= 22 \times 10^6 \text{m}^3$ gases $= 23.48 \times 10^6 \text{m}^3$

Coke $= 11.1$ tons

Ca $CO_3 = 2.41$ tons

Illustration Material Balance

Pig iron and coke are charged in a cupola to produce an iron casting. The flux is pure Ca CO_3 and Kg is used/ ton of pig iron charged.The coke used 125kg / ton of pig iron; the composition of coke is $85\%C, 6\%SiO_2 6\%Al_2O_3$ and $3\%FeO$

The gas from cupola contains $CO:CO_2 = 1:1$ by volume No carbon is oxidized from pig iron. The slag from cupola is:

$$FeO\ 12\%\ SiO_2\ 45\%MnO\ 3\%,\ CaO\ 25\%,\ Al_2O_3\ 15\%$$

The cast iron produced carries 3.8% C, besides some Mn and S.

Required Per ton of Pig Iron

a) Weight of slag

b) Volume of air consumed in oxidizing Si Mn and Fe

c) Volume of air oxidation of C of coke.

d) Volume and % composition of gas

Solution

Al_2O_3 balance given weight of slag $=50$ kg

To find amount of air, we have to find oxygen of air used to oxidize Si, Mn and Fe.

Si oxidized $=$ Si in ch arg e $-$ Si in slag

O_2 required for Si oxidation $= 0.25$ kg mole

O_2 required for Mn oxidation $= 0.011$ kg mole

O_2 required for Fe oxidation $= 0.0156$ kg mole

Volume of air consumed in oxidation $= 29.504 m^3$

Volume of air for oxidation of carbon of coke can be found one we know CO and CO_2. Since C oxidizes to CO and CO_2.

Let Y kg mole $CO + CO_2$ in exit gas

$\therefore$ exit gas $: 0.5Y$ kg mole CO and $0.5Y$ kg mole CO_2

Carbon and oxygen balance: Let x kg mole of oxygen is required for oxidation of carbon of coke

$$8.854 + 0.26 = 0.5Y + 0.5Y$$

$$x + 0.411 = 0.75$$

Solving Above Equations we Get

$$x = 6.4245 \text{ and } Y = 9.114$$

Volume of air $= 685.28 \text{ m}^3$

Volume of gas $= 768.5 \text{m}^3$

$CO = 13.28\%, CO_2 = 13.28\% \ N_2 = 73.44\%$

Gaseous Fuel

Gaseous fuel is produced by gasifying coal or coke in a reactor called gas producer. Gaseous fuels have several advantages like

- It is easy to handle.

- Combustion is rapid in comparison to coal or fuel oil.

- Less excess air for combustion is required than for combustion of fuel oil and coke.

Thermodynamics of Gasification:

Thermodynamics deals with the conversion of carbon of fuel to gaseous product at equilibrium, which means thermodynamics deals with initial and final states.

Consider gasification of 1 mole of carbon with air. Stoichiometrically ½ mole of oxygen is required to produce 1 mole of CO. One mole of oxygen is obtained from 4.76 moles of air, which means that every mole of oxygen carries 3.76 moles of N_2 with it. Thus if the initial state of reactants is C, O_2 and and final state is CO and N_2 the following gasification reaction can be written:

$$C + \tfrac{1}{2}(O2 + 3.76N_2) = CO + 1.88N_2$$

(1) $-\Delta H^{\circ}_{fCO} = 29.6 \times 10^3 \text{ Kcal / kg mol}$

Gaseous fuel produced by gasification of carbon consists of CO and N_2 in which

$\%CO = 34.7\%$ and $\%N_2 = 65.3\%$

Volume of gaseous fuel $= 5.38 \text{m}^3$ / kg mole of carbon (at 273K and 1 atm)

One can also use a mixture of air + steam to gasify the fuel. Now suppose we gasify carbon with a mixture of air + steam. Note that reaction 1 generates $= 29.6 \times 10^3 / 12 = 2467$ Kcal of heat per Kg of carbon. This excessive amount of heat can generate a very high temperature in the gasifier, if the excessive amount of heat is not properly managed. In large sized gas producers heat losses are very small and there occurs substantial rise in temperature. Steam is utilized to use the heat produced by reaction 1. Steam usage brings the following advantages:

- Decomposition of steam produces hydrogen and thereby producer gas is enriched in calorific value.

- Gaseous fuel is enriched per unit volume since volume of H_2 << volume of N_2

- Excessive heat in the producer is utilized since decomposition of steam is endothermic.

Consider the following gasification reaction

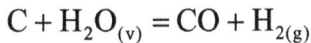

$$C + H_2O_{(v)} = CO + H_{2(g)}$$

Heat of reaction $\Delta H_R^{\circ} = (-29.6 \times 10^3 + 57.8 \times 10^3) = 28.2 \times 10^3 \, \text{Kcal}$

In the gasification of coal or coke with a mixture of air and steam, it is important to know how much amount of steam can be fed without supplying any heat from outside. Heat balance can yield the amount of steam which can be fed.

Consider gasification of 1 Kg mole of C with a mixture air and steam under adiabatic conditions:

Let X Kg mole of C reacts with air. Assuming that all steam decomposes to H_2 and carbon forms CO.

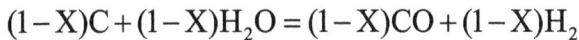

$$XC + X/2(O_2 + 3.76 \, N_2) = XCO + 1.88X \, N_2$$

$$(1-X)C + (1-X)H_2O = (1-X)CO + (1-X)H_2$$

Heat produced by above 1st eq. = Heat consumed by above 2nd eq.

$$29.6 \times 10^3 X = (1-X) \times 10^3 \times 28.2X = 0.488 \, \text{Kg mole}$$

Therefore final equations for gasification becomes

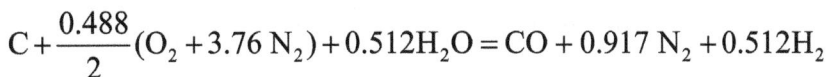

$$C + \frac{0.488}{2}(O_2 + 3.76 \, N_2) + 0.512H_2O = CO + 0.917 \, N_2 + 0.512H_2$$

Fuel gas analysis is $CO = 41.1\% \, N_2 = 37.8\%$ and $= 21.1\%$

Amount of fuel gas $= (1 + 0.917 + 0.512) \times 22.4 / 12 = 4.53 \text{ m}^3 / \text{Kg Carbon (at 273 K and 1 atm)}$

Amount of steam is 0.768 Kg/kg of carbon

Alternatively a mixture of oxygen and steam can also be used for gasification. Consider the gasification of a mole of carbon by a mixture of pure oxygen +steam

Assuming that all steam decomposes to H_2 and carbon forms CO.

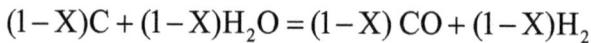

$$XC + X/2O_2 + = XCO$$

$$(1-X)C + (1-X)H_2O = (1-X) CO + (1-X)H_2$$

Heat produced by above 1st eq. = Heat consumed by above 2nd eq.

$$29.6 \times 10^3 X = (1-X) \times 10^3 \times 28.2$$

$$X = 0.488 \text{ Kg mole}$$

Therefore final equations for gasification becomes

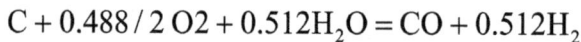

$$C + 0.488/2 \text{ O2} + 0.512H_2O = CO + 0.512H_2$$

Fuel gas analysis is CO $= 66.14\%$, and $H_2 = 33.86\%$

Amount of fuel gas $= 2.82 \text{ m}^3 / \text{Kg Carbon (at 273 K and 1 atm)}$

Amount of steam is 0.768 Kg/kg of carbon

Calorific Value

Calorific vale is the potential enrgy contained in the fuel. The calculation of calorific value is illustrated in the following on the basis of gasification of 1 kg of carbon. The calorific value of gaseous fuel when air is used for gasification is 5633kcal

Calorific Value of C = 8100 Kcal calculated by Dulong's formula using pure C

Therefore % CV of 1kg of C available in producer gas is 69.5%

That means 30.5% of calorific value of C represents sensible heat and heat losses.

Calorific value of producer gas when produced by gasifying carbon with a mixture of air+steam is

$= 8077$ kcal when standard state of the product is gas.

Calorific value of producer gas as expressed in % of calorific value of 1 kg of $C = 98.8\%$

We observe that the gasification of C with a mixture of air and steam increases the calorific value of producer gas which is mainly due to addition of hydrogen.

The calorific value of producer gas obtained by a mixture of oxygen and steam is almost similar to that of air + steam. But the volume of producer gas is only 62% that of air + steam and 48% that of air only. Benefit of using steam is self evident.

Is it Possible to Decompose All Steam to Hydrogen and C to CO?

Typically, gas producer, operates in a counter-current mode i.e. coal is charged from top and a mixture of air and steam is blown through the coal bed simultaneously and continuously from the bottom. The flow rates of air and steam are adjusted so that the heat evolution in the reaction of with C of coke/coal balances the heat absorption due to endothermic reaction of decomposition of steam. Thus

- Not all steam decomposes to H_2 i.e. some amount of steam remains undecomposed and

- Not all C is converted to CO.

The extent of decomposition of steam to hydrogen and conversion of C to CO depends on temperature, residence time of the reactants in the reactor, reactivity of carbon of coal, reaction surface area and etc. The gaseous fuel produced after gasification with a moisture of air +s team and carbon will always contain undecomposed steam and carbon dioxide besides other components.

Basics

The following diagram shows the input of materials like coal, air and steam in a gasifier and the outputs are producer gas, ashes, tar and soot.

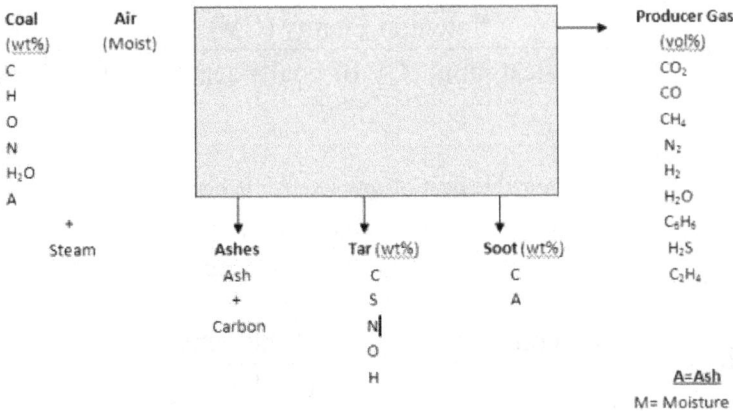

Material balance in gasification

Basis: 1000 Kg Coal

a) Amount of producer gas

Carbon balance

C from coal = C in ashes + C in tar + C in soot + C in producer gas

b) To calculate amount of steam decomposed

Decomposition of steam produces H_2.

Moisture of coal directly enters into PG without being decomposed.

Moisture of air and steam decompose to H_2 and is included in CH_4, H_2 and other hydrocarbons.

H Balance

H from coal+ H from Moisture of coal + H from steam + H from moist air = H in tar + H in PG (producer gas)

c) Water in producer gas = Moisture from coal + undecomposed steam

d) Nitrogen balance for amount of air

Oxygen balance if required to check the results of calculation.

e) Ash balance to know amount of ashes, if not given.

The raw hot gas from producer can be delivered through insulated mains as such to the furnaces and plants nearby. So that both potential energy of gas (CV) and sensible heat can be utilized. A more prevalent practisc is to cool the gas and purify it to remove deleterious constituents, for example H2S and then distribute to plants.

$$\text{Cold gas efficiency} = \frac{\text{Potential Energy (CV) of gas made} \times 100}{\text{Total heat input (CV of coal + sensible heat of coal, air, steam)}}$$

Hot gas efficiency

$$= \frac{\begin{array}{c} \text{(PE of gas + sensible heat of gas + sensible heat of water vapour + PE of tar} \\ \text{+ PE of soot + sensible heat of tar + sensible heat of soot)} \times 100 \end{array}}{\text{Total heat input}}$$

$$\text{Thermal efficiency} = \frac{\text{(Potential energy of gas + enthalpy of steam produced)} \times 100}{\text{Total heat input}}$$

Normally following efficiency values are reported in the literature:

Cold gas efficiency ≈ 60-80%

Hot gas efficiency ≈ 90%

Losses ≈ 9%

Example

Determine Material and heat balance of a gasifier and calculate efficiencies. The analysis of various inputs and outputs are given. Temperatures of input and outputs are also given.

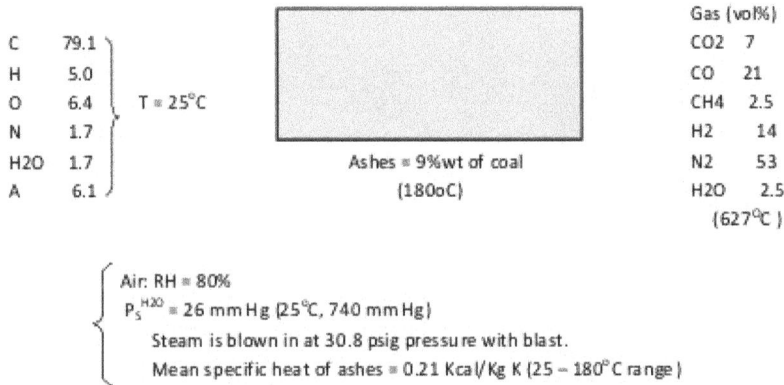

				Gas (vol%)	
C	79.1			CO2	7
H	5.0			CO	21
O	6.4	T = 25°C		CH4	2.5
N	1.7			H2	14
H2O	1.7		Ashes = 9%wt of coal	N2	53
A	6.1		(180oC)	H2O	2.5
				(627°C)	

Air: RH = 80%

P_s^{H2O} = 26 mm Hg (25°C, 740 mm Hg)

Steam is blown in at 30.8 psig pressure with blast.

Mean specific heat of ashes = 0.21 Kcal/Kg K (25 – 180°C range)

Material Balance Diagram

Basis 1 Kg coal

Volume of producer gas(fuel gas)

Let Y Kg mole producer gas

C in coal = C in producer gas + C in ashes

0.791/12 = (0.07 + 0.21 + 0.025) Y + (0.09 – 0.061)

Y = 0.208 Kg mole or = 4.66 m³/Kg coal (1 atm, 273K)

Volume of air (moist)

Let X Kg mole moist air

Since the air is moist, we have to calculate composition of air.

$P_{N2} + P_{O2} + P_{H2O}$ = 740 mm Hg

$P_{N2} + P_{O2}$ = 740 – 0.8 × 26

$P_{N2} + P_{O2}$ = 719.2 mm Hg

P_{N2} = 568.168 mm

P_{O2} = 151.032 mm

P_{H2O} = 20.800 mm

Composition of 1 Kg Mole of Moist Air

N_2 = 0.7677

O_2 = 0.2041

H_2O = 0.0281

N_2 Balance

N in coal + N_2 from moist air = N_2 in Producer gas

0.017/28 + 0.7677X = 0.53 × 0.208

X = 0.14279 Kg mole or = 3.601 m³ (26°C and 740 mm Hg)

Weight of Steam : Hydrogen Balance

Consider Z Kg mole steam.

0.025 + 0.00094 + Z + 0.00401 = 0.004472

Z = 0.015 Kg mole

= 0.266 Kg steam/Kg coal

% H2O blown in, that was decomposed

Water vapour in PG = Water from evaporation of M of coal + Water of undecomposed steam

0.025 × 0.208 = 0.017/18 + W

W = 0.004255 Kg mole undecomposed steam

Steam decomposed = {0.266 – (0.004255 × 18)}

= 0.1895 Kg

% steam blown, that is decomposed in producer gas = 0.1895 × 100/ 0.266

= 71.2

NVC pf producer gas

	Kg moles	Kg moles	
CO	0.04368	-67.6×103	
CH_4	0.0052	-194.91×103	NVC = 5.64 × 103 Kcal
H_2	0.02912	57.8×103	

	Kg moles	Kcal/Kg mole	
CO	0.04368	-67.6×103	NCV = 5.64×103 Kcal
CH$_4$	0.0052	-194.91×103	
H$_2$	0.02912	57.8×103	

NCV of Coal

$= 81 \%C + 341 [\%H - \%O/8] - 5.84 (9 \%H + M)$

$= 81 \times 79.1 + 341 [5 - 6.4/8] - 5.84 (9 \times 5 + 1.7)$

$= 7566.32$ Kcal

Enthalpy of Water Vapour in Moist Air

$H_2O_{(l)} = H_2O_{(g)}$

Heat absorbed = 584 Kcal/Kg H_2O

$\qquad = 584 \times 1.7 /100 = 9.93$ Kcal

Enthalpy of Saturated Steam

Gauge pressure = 30.8 psi

Pressure 740 mm = 14.3 psi

Absolute pressure = 45.1 psi

Enthalpy of saturated steam at 45 psi referred to water at 0°C = 651 Kcal/Kg

Enthalpy difference between water at 25°C and water at 0°C = 24.94 Kcal/Kg

Enthalpy of steam referred to water at 25°C = 626 Kcal/Kg

Enthalpy of steam used = 626 × 0.266

$\qquad = 166$ Kcal

Enthalpy of Water Vapour in Hot Gas at 900K

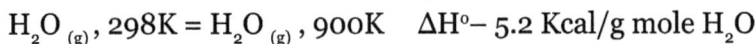

$H_2O_{(l)} = H_2O_{(g)}$ $\qquad \Delta H°_{298} = 10.5$ Kcal/g mole H_2O

$H_2O_{(g)}, 298K = H_2O_{(g)}, 900K$ $\quad \Delta H° - 5.2$ Kcal/g mole H_2O

Enthalpy of water vapour referred to $H_2O_{(l)}$ = 15.7 Kcal/g mole H_2O

Enthalpy of water vapour in hot gas = 15.7 × 0.208 × 1000 × 2.5/100

$\qquad = 81.64$ Kcal

Sensible heat of dry producer gas at 900K

$H_{900} - H_{298} \mid CO_2 = 6708$ Kcal/Kg mole

$H_{900} - H_{298} \mid CO = 4400$ Kcal/Kg mole

$H_{900} - H_{298} \mid CH_4 = 7522$ Kcal/Kg mole

$H_{900} - H_{298} \mid H_2 = 4224$ Kcal/Kg mole

$H_{900} - H_{298} \mid N_2 = 4358$ Kcal/Kg mole

Heat Balance

Heat Input

Input	Kcal
CV of coal	7566.32
Sensible heat in coal, air	0
Enthalpy of water vapour in air	9.93
Enthalpy of steam	166
Total	7742.25

Heat Output

Output	Kcal
CV of dry PG	5640
Sensible heat of dry PG	932.8
Enthalpy of water vapour in hot gas	81.6
Heat losses	1087.85
Total	7742.25

Cold gas efficiency = 5640 ×100/7742.25

$\qquad$ = 72.85%

Hot Gas efficiency = 6653.6 × 100/7742.25

$\qquad$ = 85.9 %

Thermal efficiency = 5721.6 × 100/7742.25

$\qquad$ = 73.9%

Mechanics of Furnace

A furnace is essentially a thermal enclosure and is employed to process raw materials at high temperatures both in solid state and liquid state. Several industries like iron and steel making, non ferrous metals production, glass making, manufacturing, ceramic processing, calcination in cement production etc. employ furnace. The principle objectives are

a) To utilize heat efficiently so that losses are minimum, and

b) To handle the different phases (solid, liquid or gaseous) moving at different velocities for different times and temperatures such that erosion and corrosion of the refractory are minimum.

Source of Energy

- Combustion of fossil fuels, that is solid, liquid and gaseous.

- Electric energy: Resistance heating, induction heating or arc heating.

- Chemical energy: Exothermic reactions

Types of Furnaces

Furnaces are both batch and continuous type. In the continuous type for example in heating of ferrous material for hot working, the furnace chamber consists of preheating, heating and soaking zones. The material enters through the preheating zone and exits the soaking zone for rolling. But the flow of products of combustion is in the reverse direction. Furnace design is recuperative type in that material exits at the desired temperature from the soaking zone and the products of combustion discharge the preheating zone at the lowest possible temperature. Different types of continuous furnaces are in use, like walking beam type, pusher type, roller hearth type, screw conveyor type etc.

In the batch furnaces, the load is heated for the fixed time and then discharged from the furnace. There are different types of batch furnaces like box type, integral quench type, pit type and car.

In many cases the furnace is equipped with either external heat recovery system or internal heat recovery system. In the external heat recovery system a heat exchanger like recuperator is installed outside the furnace. Here heat exchanger must be integrated with the furnace operation. In the internal heat recovery the products of combustion are recirculated in the furnace itself so that flame temperature is some what lowered. The objective is to reduce the NOx formation ottom type.

Industrial Furnace

An Industrial Furnace

The purpose of an industrial furnace is to attain a higher processing temperature in comparison to open-air systems, as well as the efficiency gains of a closed system. Industrial furnaces typically deal with temperatures higher than 400 degrees Celsius.

An industrial furnace is an equipment used to provide heat for a process or can serve as reactor which provides heats of reaction. Furnace designs vary as to its function, heating duty, type of fuel and method of introducing combustion air.

Heat is generated by an industrial furnace by mixing fuel with air or oxygen, or from electrical energy. The residual heat will exit the furnace as flue gas.

Overview

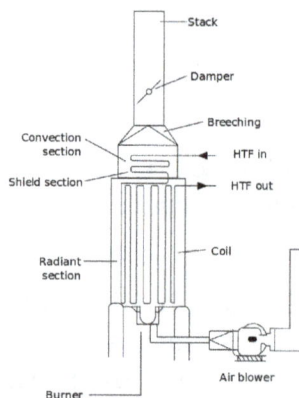

Schematic diagram of an industrial process furnace

Fuel flows into the burner and is burnt with air provided from an air blower. There can be more than one burner in a particular furnace which can be arranged in cells which heat a particular set of tubes. Burners can also be floor mounted, wall mounted or roof mounted depending on design. The flames heat up the tubes, which in turn heat the fluid inside in the first part of the furnace known as the radiant section or firebox. In this chamber where combustion takes place, the heat is transferred mainly by radiation to tubes around the fire in the chamber.

The heating fluid passes through the tubes and is thus heated to the desired temperature. The gases from the combustion are known as flue gas. After the flue gas leaves the firebox, most furnace designs include a convection section where more heat is recovered before venting to the atmosphere through the flue gas stack. (HTF=Heat Transfer Fluid. Industries commonly use their furnaces to heat a secondary fluid with special additives like anti-rust and high heat transfer efficiency. This heated fluid is then circulated round the whole plant to heat exchangers to be used wherever heat is needed instead of directly heating the product line as the product or material may be volatile or prone to cracking at the furnace temperature.)

Components

Radiant Section

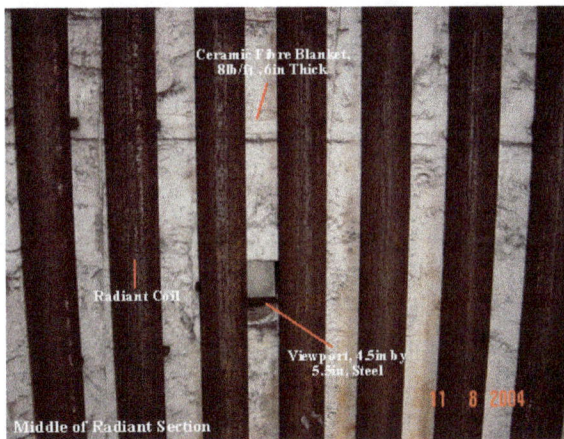

Middle of radiant section

The radiant section is where the tubes receive almost all its heat by radiation from the flame. In a vertical, cylindrical furnace, the tubes are vertical. Tubes can be vertical or horizontal, placed along the refractory wall, in the middle, etc., or arranged in cells. Studs are used to hold the insulation together and on the wall of the furnace. They are placed about 1 ft (300 mm) apart in this picture of the inside of a furnace.

The tubes, shown below, which are reddish brown from corrosion, are carbon steel tubes and run the height of the radiant section. The tubes are a distance away from the insulation so radiation can be reflected to the back of the tubes to maintain a uniform tube wall temperature. Tube guides at the top, middle and bottom hold the tubes in place.

Convection Section

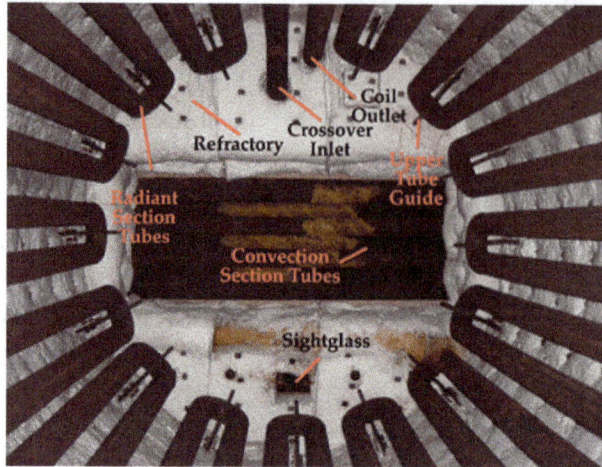

Convection section

The convection section is located above the radiant section where it is cooler to recover additional heat. Heat transfer takes place by convection here, and the tubes are finned to increase heat transfer. The first two tube rows in the bottom of the convection section and at the top of the radiant section is an area of bare tubes (without fins) and are known as the shield section ("shock tubes"), so named because they are still exposed to plenty of radiation from the firebox and they also act to shield the convection section tubes, which are normally of less resistant material from the high temperatures in the firebox.

The area of the radiant section just before flue gas enters the shield section and into the convection section called the bridgezone. A crossover is the tube that connects from the convection section outlet to the radiant section inlet. The crossover piping is normally located outside so that the temperature can be monitored and the efficiency of the convection section can be calculated. The sightglass at the top allows personnel to see the flame shape and pattern from above and visually inspect if flame impingement is occurring. Flame impingement happens when the flame touches the tubes and causes small isolated spots of very high temperature.

Radiant Coil

This is a series of tubes horizontal/ vertical hairpin type connected at ends (with 180° bends) or helical in construction. The radiant coil absorbs heat through radiation. They can be single pass or multi pass depending upon the process-side pressure drop allowed. The radiant coils and bends are housed in the radiant box. Radiant coil materials vary from carbon steel for low temperature services to high alloy steels for high temperature services. These are supported from the radiant side walls or hanging from the radiant roof. Material of these supports is generally high alloy steel. While designing the radiant coil, care is taken so that provision for expansion (in hot conditions) is kept.

Burner

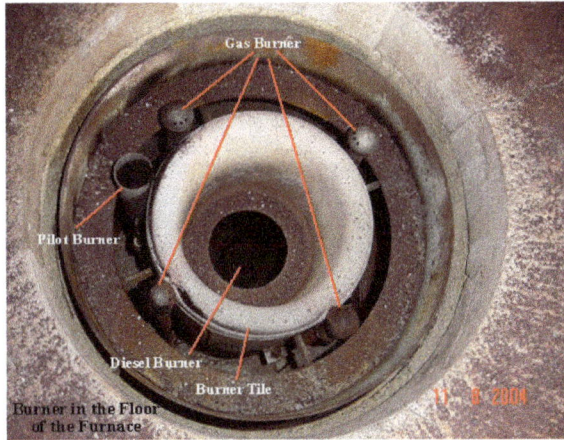

Furnace burner

The burner in the vertical, cylindrical furnace as above, is located in the floor and fires upward. Some furnaces have side fired burners, such as in train locomotives. The burner tile is made of high temperature refractory and is where the flame is contained. Air registers located below the burner and at the outlet of the air blower are devices with movable flaps or vanes that control the shape and pattern of the flame, whether it spreads out or even swirls around. Flames should not spread out too much, as this will cause flame impingement. Air registers can be classified as primary, secondary and if applicable, tertiary, depending on when their air is introduced.

The primary air register supplies primary air, which is the first to be introduced in the burner. Secondary air is added to supplement primary air. Burners may include a pre-mixer to mix the air and fuel for better combustion before introducing into the burner. Some burners even use steam as premix to preheat the air and create better mixing of the fuel and heated air. The floor of the furnace is mostly made of a different material from that of the wall, typically hard castable refractory to allow technicians to walk on its floor during maintenance.

A furnace can be lit by a small pilot flame or in some older models, by hand. Most pilot flames nowadays are lit by an ignition transformer (much like a car's spark plugs). The pilot flame in turn lights up the main flame. The pilot flame uses natural gas while the main flame can use both diesel and natural gas. When using liquid fuels, an atomizer is used, otherwise, the liquid fuel will simply pour onto the furnace floor and become a hazard. Using a pilot flame for lighting the furnace increases safety and ease compared to using a manual ignition method (like a match).

Sootblower

Sootblowers are found in the convection section. Sootblowing is normally done when the efficiency of the convection section is decreased. This can be calculated by looking at the temperature change from the crossover piping and at the convection section exit.

Sootblowers utilize flowing media such as water, air or steam to remove deposits from the tubes. This is typically done during maintenance with the air blower turned on. There are several different types of sootblowers used. Wall blowers of the rotary type are mounted on furnace walls protruding between the convection tubes. The lances are connected to a steam source with holes drilled into it at intervals along its length. When it is turned on, it rotates and blows the soot off the tubes and out through the stack.

Stack

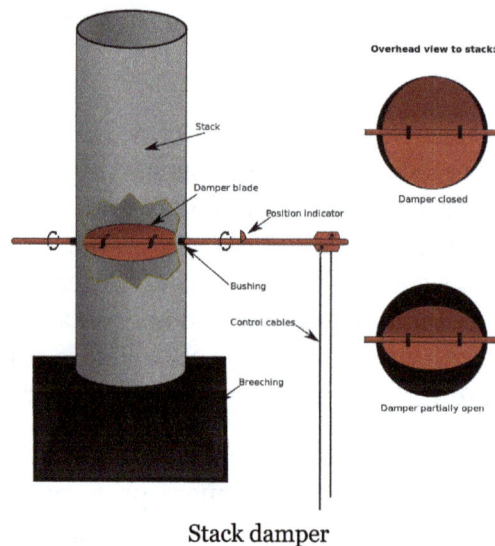

Stack damper

The flue gas stack is a cylindrical structure at the top of all the heat transfer chambers. The breeching directly below it collects the flue gas and brings it up high into the atmosphere where it will not endanger personnel.

The stack damper contained within works like a butterfly valve and regulates draft (pressure difference between air intake and air exit) in the furnace, which is what pulls the flue gas through the convection section. The stack damper also regulates the heat lost through the stack. As the damper closes, the amount of heat escaping the furnace through the stack decreases, but the pressure or draft in the furnace increases which poses risks to those working around it if there are air leakages in the furnace, the flames can then escape out of the firebox or even explode if the pressure is too great.

Insulation

Insulation is an important part of the furnace because it improves efficiency by minimizing heat escape from the heated chamber. Refractory materials such as firebrick, castable refractories and ceramic fibre, are used for insulation. The floor of the furnace are normally castable type refractories while those on the walls are nailed or glued in place. Ceramic fibre is commonly used for the roof and wall of the furnace and is graded by its density and then its maximum temperature rating. For example, 8# 2,300 °F means 8 lb/ft³ density with a maximum temperature rating of 2,300 °F. The actual service temperature rating for ceramic fiber is a bit lower than the maximum rated temperature. (i.e. 2300 °F is only good to 2145 °F before permanent linear shrinkage).

Foundations

Concrete pillars are foundation on which the heater is mounted. They can be four nos. for smaller heaters and may be up to 24 nos. for large size heaters. Design of pillars and entire foundation is done based on the load bearing capacity of soil and seismic conditions prevailing in the area. Foundation bolts are grouted in foundation after installation of the heater.

Access Doors

The heater body is provided with access doors at various locations. Access doors are to be used only during shutdown of heater. The normal size of the access door is 600x400 mm, which is sufficient for movement of men/ material into and out of the heater. During operation the access doors are properly bolted using leak proof high temperature gaskets.

Electric arc Furnace

An electric arc furnace (the large cylinder) being tapped

An electric arc furnace (EAF) is a furnace that heats charged material by means of an electric arc.

Rendering of exterior and interior of an electric arc furnace

Industrial arc furnaces range in size from small units of approximately one ton capacity (used in foundries for producing cast iron products) up to about 400 ton units used for secondary steelmaking. Arc furnaces used in research laboratories and by dentists may have a capacity of only a few dozen grams. Industrial electric arc furnace temperatures can be up to 1,800 °C (3,272 °F), while laboratory units can exceed 3,000 °C (5,432 °F).

Arc furnaces differ from induction furnaces in that the charge material is directly exposed to an electric arc, and the current in the furnace terminals passes through the charged material.

History

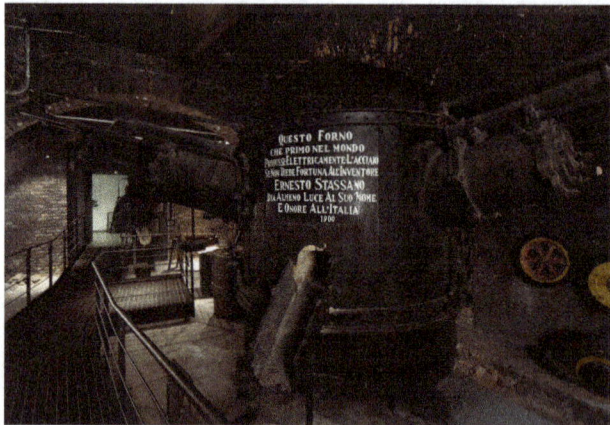

Stassano furnace exhibited at the Museo della Scienza e della Tecnologia "Leonardo da Vinci", Milan

In the 19th century, a number of men had employed an electric arc to melt iron. Sir Humphry Davy conducted an experimental demonstration in 1810; welding was investigated by Pepys in 1815; Pinchon attempted to create an electrothermic furnace in 1853; and, in 1878–79, Sir William Siemens took out patents for electric furnaces of the arc type.

The first electric arc furnaces were developed by Paul Héroult, of France, with a commercial plant established in the United States in 1907. The Sanderson brothers formed

The Sanderson Brothers steel Co. in Syracuse, New York, installing the first electric arc furnace in the U.S. This furnace is now on display at Station Square, Pittsburgh, Pennsylvania.

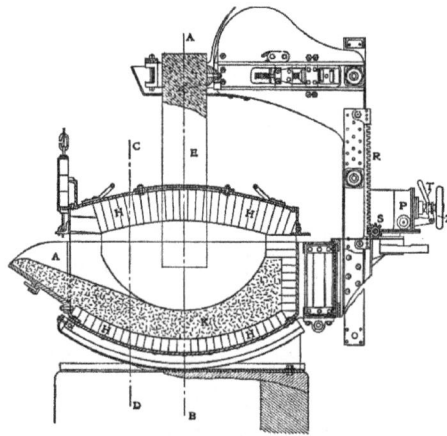

Transvers Section E F
FIG. 306. — HÉROULT REFINING FURNACE.

A schematic cross section through a Heroult arc furnace. E is an electrode (only one shown), raised and lowered by the rack and pinion drive R and S. The interior is lined with refractory brick H, and K denotes the bottom lining. A door at A allows access to the interior. The furnace shell rests on rockers to allow it to be tilted for tapping.

Initially "electric steel" was a specialty product for such uses as machine tools and spring steel. Arc furnaces were also used to prepare calcium carbide for use in carbide lamps. The *Stassano electric furnace* is an arc type furnace that usually rotates to mix the bath. The *Girod furnace* is similar to the *Héroult furnace*.

While EAFs were widely used in World War II for production of alloy steels, it was only later that electric steelmaking began to expand. The low capital cost for a mini-mill— around US$140–200 per ton of annual installed capacity, compared with US$1,000 per ton of annual installed capacity for an integrated steel mill—allowed mills to be quickly established in war-ravaged Europe, and also allowed them to successfully compete with the big United States steelmakers, such as Bethlehem Steel and U.S. Steel, for low-cost, carbon steel "long products" (structural steel, rod and bar, wire, and fasteners) in the U.S. market.

When Nucor—now one of the largest steel producers in the U.S.—decided to enter the long products market in 1969, they chose to start up a mini-mill, with an EAF as its steelmaking furnace, soon followed by other manufacturers. Whilst Nucor expanded rapidly in the Eastern US, the companies that followed them into mini-mill operations concentrated on local markets for long products, where the use of an EAF allowed the plants to vary production according to local demand. This pattern was also followed globally, with EAF steel production primarily used for long products, while integrated mills, using blast furnaces and basic oxygen furnaces, cornered the markets for "flat products"—sheet steel and heavier steel plate. In 1987, Nucor made the decision to expand into the flat products market, still using the EAF production method.

Construction

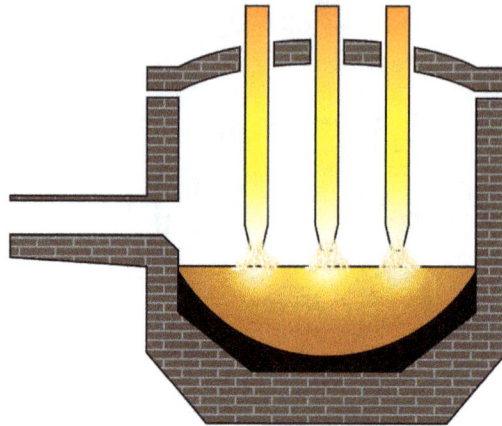

A schematic cross-section through an EAF. Three electrodes (yellow), molten bath (gold), tapping spout at left, refractory brick movable roof, brick shell, and a refractory-lined bowl-shaped hearth

An electric arc furnace used for steelmaking consists of a refractory-lined vessel, usually water-cooled in larger sizes, covered with a retractable roof, and through which one or more graphite electrodes enter the furnace. The furnace is primarily split into three sections:

- the *shell*, which consists of the sidewalls and lower steel "bowl";

- the *hearth*, which consists of the refractory that lines the lower bowl;

- the *roof*, which may be refractory-lined or water-cooled, and can be shaped as a section of a sphere, or as a frustum (conical section). The roof also supports the refractory delta in its centre, through which one or more graphite electrodes enter.

The hearth may be hemispherical in shape, or in an eccentric bottom tapping furnace, the hearth has the shape of a halved egg. In modern meltshops, the furnace is often raised off the ground floor, so that ladles and slag pots can easily be manufactured under either end of the furnace. Separate from the furnace structure is the electrode support and electrical system, and the tilting platform on which the furnace rests. Two configurations are possible: the electrode supports and the roof tilt with the furnace, or are fixed to the raised platform.

A typical alternating current furnace is powered by a three-phase electrical supply and therefore has three electrodes. Electrodes are round in section, and typically in segments with threaded couplings, so that as the electrodes wear, new segments can be added. The arc forms between the charged material and the electrode, the charge is heated both by current passing through the charge and by the radiant energy evolved by the arc. The electric arc temperature reaches around 3000 °C (5000 °F), thus causing the lower sections of the electrodes glow incandescently when in operation. The electrodes are automatically raised and lowered by a positioning system, which may use either electric winch hoists or hydraulic cylinders. The regulating system maintains approximately constant current and power input during the melting of the charge, even though scrap may move under the electrodes as it melts. The mast arms holding the electrodes can either carry heavy busbars (which may be hollow water-cooled copper pipes carrying current to the electrode clamps) or be "hot arms", where the whole arm carries the current, increasing efficiency. Hot arms can be made from copper-clad steel or aluminium. Since the electrodes move up and down automatically for regulation of the arc, and are raised to allow removal of the furnace roof, large water-cooled cables connect the bus tubes/arms with the transformer located adjacent to the furnace. To protect the transformer from heat, it is installed in a vault and is itself cooled via pumped oil exchanging heat with the plant's water-cooling systems, as the electrical conditions for arc-furnace steelmaking are extremely stressful on the transformer.

The roof of an arc furnace removed, showing the three electrodes

The furnace is built on a tilting platform so that the liquid steel can be poured into another vessel for transport. The operation of tilting the furnace to pour molten steel is called "tapping". Originally, all steelmaking furnaces had a tapping spout closed with refractory that washed out when the furnace was tilted, but often modern furnaces have an eccentric bottom tap-hole (EBT) to reduce inclusion of nitrogen and slag in the liquid steel. These furnaces have a taphole that passes vertically through the hearth and shell, and is set off-centre in the narrow "nose" of the egg-shaped hearth. It is filled with refractory sand, such as olivine, when it is closed off. Modern plants may have two shells with a single set of electrodes that can be transferred between the two; one shell preheats scrap while the other shell is utilised for meltdown. Other DC-based furnaces have a similar arrangement, but have electrodes for each shell and one set of electronics.

AC furnaces usually exhibit a pattern of hot and cold-spots around the hearth perimeter, with the cold-spots located between the electrodes. Modern furnaces mount oxygen-fuel burners in the sidewall and use them to provide chemical energy to the cold-spots, making the heating of the steel more uniform. Additional chemical energy is provided by injecting oxygen and carbon into the furnace; historically this was done through lances (hollow mild-steel tubes) in the slag door, now this is mainly done through wall-mounted injection units that combine the oxygen-fuel burners and the oxygen or carbon injection systems into one unit.

A mid-sized modern steelmaking furnace would have a transformer rated about 60,000,000 volt-amperes (60 MVA), with a secondary voltage between 400 and 900 volts and a secondary current in excess of 44,000 amperes. In a modern shop such a furnace would be expected to produce a quantity of 80 metric tonnes of liquid steel in approximately 50 minutes from charging with cold scrap to tapping the furnace. In comparison, basic oxygen furnaces can have a capacity of 150–300 tonnes per batch, or "heat", and can produce a heat in 30–40 minutes. Enormous variations exist in furnace design details and operation, depending on the end product and local conditions, as well as ongoing research to improve furnace efficiency. The largest scrap-only furnace (in terms of tapping weight and transformer rating) is a DC furnace operated by Tokyo Steel in Japan, with a tap weight of 420 metric tonnes and fed by eight 32MVA transformers for 256MVA total power.

To produce a ton of steel in an electric arc furnace requires approximately 400 kilowatt-hours per short ton or about 440 kWh per metric tonne; the theoretical minimum amount of energy required to melt a tonne of scrap steel is 300 kWh (melting point 1520 °C/2768 °F). Therefore, a 300-tonne, 300 MVA EAF will require approximately 132 MWh of energy to melt the steel, and a "power-on time" (the time that steel is being melted with an arc) of approximately 37 minutes. Electric arc steelmaking is only economical where there is plentiful electricity, with a well-developed electrical grid. In many locations, mills operate during off-peak hours when utilities have surplus power generating capacity and the price of electricity is less.

Operation

Scrap metal is delivered to a scrap bay, located next to the melt shop. Scrap generally comes in two main grades: shred (whitegoods, cars and other objects made of similar light-gauge steel) and heavy melt (large slabs and beams), along with some direct reduced iron (DRI) or pig iron for chemical balance. Some furnaces melt almost 100% DRI.

An arc furnace pouring out steel into a small ladle car. For scale, note the operator standing on the platform at upper left. This is a 1941-era photograph and so does not have the extensive dust collection system that a modern installation would have, nor is the operator wearing a hard hat or dust mask.

The scrap is loaded into large buckets called baskets, with "clamshell" doors for a base. Care is taken to layer the scrap in the basket to ensure good furnace operation; heavy melt is placed on top of a light layer of protective shred, on top of which is placed more shred. These layers should be present in the furnace after charging. After loading, the basket may pass to a scrap pre-heater, which uses hot furnace off-gases to heat the scrap and recover energy, increasing plant efficiency.

The scrap basket is then taken to the melt shop, the roof is swung off the furnace, and the furnace is charged with scrap from the basket. Charging is one of the more dangerous operations for the EAF operators. A lot of potential energy is released by the tonnes of falling metal; any liquid metal in the furnace is often displaced upwards and outwards by the solid scrap, and the grease and dust on the scrap is ignited if the furnace is hot, resulting in a fireball erupting. In some twin-shell furnaces, the scrap is charged into the second shell while the first is being melted down, and pre-heated with off-gas from the active shell. Other operations are continuous charging—pre-heating scrap on a conveyor belt, which then discharges the scrap into the furnace proper, or charging the scrap from a shaft set above the furnace, with off-gases directed through the shaft. Other furnaces can be charged with hot (molten) metal from other operations.

After charging, the roof is swung back over the furnace and meltdown commences. The electrodes are lowered onto the scrap, an arc is struck and the electrodes are then set to bore into the layer of shred at the top of the furnace. Lower voltages are selected for this first part of the operation to protect the roof and walls from excessive heat and damage from the arcs. Once the electrodes have reached the heavy melt at the base of the furnace and the arcs are shielded by the scrap, the voltage can be increased and the electrodes raised slightly, lengthening the arcs and increasing power to the melt. This enables a molten pool to form more rapidly, reducing tap-to-tap times. Oxygen is blown into the scrap, combusting or cutting the steel, and extra chemical heat is provided by wall-mounted oxygen-fuel burners. Both processes accelerate scrap meltdown. Supersonic nozzles enable oxygen jets to penetrate foaming slag and reach the liquid bath.

An important part of steelmaking is the formation of slag, which floats on the surface of the molten steel. Slag usually consists of metal oxides, and acts as a destination for oxidised impurities, as a thermal blanket (stopping excessive heat loss) and helping to reduce erosion of the refractory lining. For a furnace with basic refractories, which includes most carbon steel-producing furnaces, the usual slag formers are calcium oxide (CaO, in the form of burnt lime) and magnesium oxide (MgO, in the form of dolomite and magnesite). These slag formers are either charged with the scrap, or blown into the furnace during meltdown. Another major component of EAF slag is iron oxide from steel combusting with the injected oxygen. Later in the heat, carbon (in the form of coke or coal) is injected into this slag layer, reacting with the iron oxide to form metallic iron and carbon monoxide gas, which then causes the slag to foam, allowing greater thermal efficiency, and better arc stability and electrical efficiency. The slag blanket also covers the arcs, preventing damage to the furnace roof and sidewalls from radiant heat.

Once the scrap has completely melted down and a flat bath is reached, another bucket of scrap can be charged into the furnace and melted down, although EAF development is moving towards single-charge designs. After the second charge is completely melted, refining operations take place to check and correct the steel chemistry and superheat the melt above its freezing temperature in preparation for tapping. More slag formers are introduced and more oxygen is blown into the bath, burning out impurities such as silicon, sulfur, phosphorus, aluminium, manganese, and calcium, and removing their oxides to the slag. Removal of carbon takes place after these elements have burnt out first, as they have a greater affinity for oxygen. Metals that have a poorer affinity for oxygen than iron, such as nickel and copper, cannot be removed through oxidation and must be controlled through scrap chemistry alone, such as introducing the direct reduced iron and pig iron mentioned earlier. A foaming slag is maintained throughout, and often overflows the furnace to pour out of the slag door into the slag pit. Temperature sampling and chemical sampling take place via automatic lances. Oxygen and carbon can be automatically measured via special probes that dip into the steel, but for all other elements, a "chill" sample—a small, solidified sample of the steel—is analysed on an arc-emission spectrometer.

Once the temperature and chemistry are correct, the steel is tapped out into a preheated ladle through tilting the furnace. For plain-carbon steel furnaces, as soon as slag is detected during tapping the furnace is rapidly tilted back towards the deslagging side, minimising slag carryover into the ladle. For some special steel grades, including stainless steel, the slag is poured into the ladle as well, to be treated at the ladle furnace to recover valuable alloying elements. During tapping some alloy additions are introduced into the metal stream, and more lime is added on top of the ladle to begin building a new slag layer. Often, a few tonnes of liquid steel and slag is left in the furnace in order to form a "hot heel", which helps preheat the next charge of scrap and accelerate its meltdown. During and after tapping, the furnace is "turned around": the slag door is cleaned of solidified slag, the visible refractories are inspected and water-cooled components checked for leaks, and electrodes are inspected for damage or lengthened through the addition of new segments; the taphole is filled with sand at the completion of tapping. For a 90-tonne, medium-power furnace, the whole process will usually take about 60–70 minutes from the tapping of one heat to the tapping of the next (the tap-to-tap time).

The furnace is completely emptied of steel and slag on a regular basis so that an inspection of the refractories can be made and larger repairs made if necessary. As the refractories are often made from calcined carbonates, they are extremely susceptible to hydration from water, so any suspected leaks from water-cooled components are treated extremely seriously, beyond the immediate concern of potential steam explosions. Excessive refractory wear can lead to breakouts, where the liquid metal and slag penetrate the refractory and furnace shell and escape into the surrounding areas.

Advantages of Electric arc Furnace for Steelmaking

The use of EAFs allows steel to be made from a 100% scrap metal feedstock. This greatly reduces the energy required to make steel when compared with primary steelmaking from ores. Another benefit is flexibility: while blast furnaces cannot vary their production by much and can remain in operation for years at a time, EAFs can be rapidly started and stopped, allowing the steel mill to vary production according to demand. Although steelmaking arc furnaces generally use scrap steel as their primary feedstock, if hot metal from a blast furnace or direct-reduced iron is available economically, these can also be used as furnace feed. As EAFs require large amounts of electrical power, many companies schedule their operations to take advantage of off peak electricity pricing.

A typical steelmaking arc furnace is the source of steel for a mini-mill, which may make bars or strip product. Mini-mills can be sited relatively near to the markets for steel products, and the transport requirements are less than for an integrated mill, which would commonly be sited near a harbour for access to shipping.

Environmental Issues

Although the modern electric arc furnace is a highly efficient recycler of steel scrap, operation of an arc furnace shop can have adverse environmental effects. Much of the capital cost of a new installation will be devoted to systems intended to reduce these effects, which include:

- Enclosures to reduce high sound levels

- Dust collector for furnace off-gas

- Slag production

- Cooling water demand

- Heavy truck traffic for scrap, materials handling, and product

- Environmental effects of electricity generation

Because of the very dynamic quality of the arc furnace load, power systems may require technical measures to maintain the quality of power for other customers; flicker and harmonic distortion are common side-effects of arc furnace operation on a power system. For this reason the power station should be located as close to the EA furnaces as possible.

Other Electric arc Furnaces

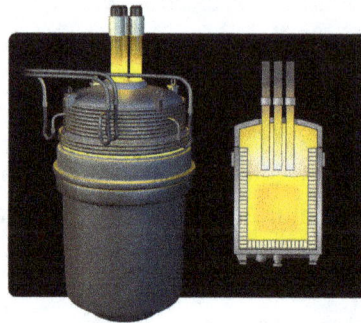

Rendering of a ladle furnace, a variation of the electric arc furnace used for keeping molten steel hot

For steelmaking, direct current (DC) arc furnaces are used, with a single electrode in the roof and the current return through a conductive bottom lining or conductive pins in the base. The advantage of DC is lower electrode consumption per ton of steel produced, since only one electrode is used, as well as less electrical harmonics and other similar problems. The size of DC arc furnaces is limited by the current carrying capacity of available electrodes, and the maximum allowable voltage. Maintenance of the conductive furnace hearth is a bottleneck in extended operation of a DC arc furnace.

In a steel plant, a ladle furnace (LF) is used to maintain the temperature of liquid steel during processing after tapping from EAF or to change the alloy composition. The ladle is used for the first purpose when there is a delay later in the steelmaking process. The ladle furnace consists of a refractory roof, a heating system, and, when applicable, a provision for injecting argon gas into the bottom of the melt for stirring. Unlike a scrap melting furnace, a ladle furnace does not have a tilting or scrap charging mechanism.

Electric arc furnaces are also used for production of calcium carbide, ferroalloys and other non-ferrous alloys, and for production of phosphorus. Furnaces for these services are physically different from steel-making furnaces and may operate on a continuous, rather than batch, basis. Continuous process furnaces may also use paste-type, Søderberg electrodes to prevent interruptions due to electrode changes. Such a furnace is known as a submerged arc furnace because the electrode tips are buried in the slag/charge, and arcing occurs through the slag, between the matte and the electrode. A steelmaking arc furnace, by comparison, arcs in the open. The key is the electrical resistance, which is what generates the heat required: the resistance in a steelmaking furnace is the atmosphere, while in a submerged-arc furnace the slag or charge forms the resistance. The liquid metal formed in either furnace is too conductive to form an effective heat-generating resistance.

Amateurs have constructed a variety of arc furnaces, often based on electric arc welding kits contained by silical blocks or flower pots. Though crude, these simple furnaces can melt a wide range of materials, create calcium carbide, etc.

Cooling Methods

Non-pressurized cooling system

Smaller arc furnaces may be adequately cooled by circulation of air over structural elements of the shell and roof, but larger installations require intensive forced cooling to maintain the structure within safe operating limits. The furnace shell and roof may be cooled either by water circulated through pipes which form a panel, or by water sprayed on the panel elements. Tubular panels may be replaced when they become cracked or reach their thermal stress life cycle. Spray cooling is the most economical and is the highest efficiency cooling method. A spray cooling piece of equipment can be relined almost endlessly; equipment that lasts 20 years is the norm. However while a tubular leak is immediately noticed in an operating furnace due to the pressure loss alarms on the panels, at this time there exists no immediate way of detecting a very small volume spray cooling leak. These typically hide behind slag coverage and can hydrate the refractory in the hearth leading to a break out of molten metal or in the worst case a steam explosion.

Plasma arc Furnace

A plasma arc furnace (PAF) uses plasma torches instead of graphite electrodes. Each of these torches consists of a casing provided with a nozzle and an axial tubing for feeding a plasma-forming gas (either nitrogen or argon), and a burnable cylindrical graphite electrode located within the tubing. Such furnaces can be referred to as "PAM" (Plasma Arc Melt) furnaces. They are used extensively in the titanium melt industry and similar specialty metals industries.

Vacuum arc Remelting

Vacuum arc remelting (VAR) is a secondary remelting process for vacuum refining and manufacturing of ingots with improved chemical and mechanical homogeneity.

In critical military and commercial aerospace applications, material engineers commonly specify VIM-VAR steels. VIM means Vacuum Induction Melted and VAR means Vacuum Arc Remelted. VIM-VAR steels become bearings for jet engines, rotor shafts for military helicopters, flap actuators for fighter jets, gears in jet or helicopter transmissions, mounts or fasteners for jet engines, jet tail hooks and other demanding applications.

Most grades of steel are melted once and are then cast or teemed into a solid form prior to extensive forging or rolling to a metallurgically sound form. In contrast, VIM-VAR steels go through two more highly purifying melts under vacuum. After melting in an electric arc furnace and alloying in an argon oxygen decarburization vessel, steels destined for vacuum remelting are cast into ingot molds. The solidified ingots then head for a vacuum induction melting furnace. This vacuum remelting process rids the steel of inclusions and unwanted gases while optimizing the chemical composition. The VIM operation returns these solid ingots to the molten state in the contaminant-free void of a vacuum. This tightly controlled melt often requires up to 24 hours. Still enveloped by the vacuum, the hot metal flows from the VIM furnace crucible into giant electrode molds. A typical electrode stands about 15 feet (5 m) tall and will be in various diameters. The electrodes solidify under vacuum.

For VIM-VAR steels, the surface of the cooled electrodes must be ground to remove surface irregularities and impurities before the next vacuum remelt. Then the ground electrode is placed in a VAR furnace. In a VAR furnace the steel gradually melts drop-by-drop in the vacuum-sealed chamber. Vacuum arc remelting further removes lingering inclusions to provide superior steel cleanliness and further remove gases such as oxygen, nitrogen and hydrogen. Controlling the rate at which these droplets form and solidify ensures a consistency of chemistry and microstructure throughout the entire VIM-VAR ingot. This in turn makes the steel more resistant to fracture or fatigue. This refinement process is essential to meet the performance characteristics of parts like a helicopter rotor shaft, a flap actuator on a military jet or a bearing in a jet engine.

For some commercial or military applications, steel alloys may go through only one vacuum remelt, namely the VAR. For example, steels for solid rocket cases, landing gears or torsion bars for fighting vehicles typically involve the one vacuum remelt.

Vacuum arc remelting is also used in production of titanium and other metals which are reactive or in which high purity is required.

How Thermal Energy is Obtained from Fossil Fuel?

All fossil fuel contain potential energy. On combustion potential energy is released in the products of combustion. The products of combustion exchange energy with the sink to raise its temperature to the required value and then exit the system. The sensible heat in POC at the critical process temperature is not available to the furnace. The higher the process critical temperature higher would be the sensible heat in POC. This sensible heat in POC is very important from the point of view of fuel utilization. We define gross available heat (GAH) as

GAH = Calorific value of fuel + sensible heat of reactants – Heat carried by POC

GAH represents the heat available at the critical process temperature; it may not represent heat available to perform a given function due to the various types of losses. GAH may be used as a criterion for comparing different fuel-combustion system.

Once the furnace is designed and built, the heat losses are not within the control of the operator; it is   governed by the process critical temperature, refractory lining thickness and thermal conductivity of the refractory. Defining net available heat (NAH) as

NAH = GAH – Heat Losses

NAH can be used as a criterion for comparing the smelting/melting/heating efficiency of different furnaces.

Variables Affecting Heat Utilization

For a given furnace design and the daily heat requirements, GAH is fixed and it is required to supply this much of heat on per day basis, we can calculate

$$\text{Fuel consumption} = \frac{\text{Required GAH per unit of time}}{\text{GAH per kg of fuel}}$$

If heat supply is the critical factor in determining the process throughput then GAH can not determine the throughput, we have to consider the NAH

$$\text{Furnace throughput} = \frac{\text{NAH generated per unit of time}}{\text{Required NAH per unit of throughput}}$$

Heat utilization or fuel utilization according to equation $m_h \times C_{ph} \, xdT_h = m_c \times C_{pc} \, xdT_c - dQ$ is inversely proportional to GAH/kg of fuel. We can derive the factors affecting heat utilization by considering 1st equation.

Air adjustment: Calorific value (CV) of fuel is the energy obtained on complete combustion of fuel with theoretical amount of air. Excess air, air leakage, furnace draft, fuel/air ratio will control the fuel consumption

Sensible heat of reactant; this heat directly adds to the furnace, fuel consumption will decrease.

POC temperature: an increase in POC temperature will increase fuel consumptionIncomplete combustion or un-burnt fuel; corresponding to incomplete combustion part of the CV of the fuel is lost with the products of incomplete combustion

Heat Utilization: Concepts

Efficient utilization of fossil fuel reserves requires, in addition to other factors, utilization of heat of POC exiting the furnace. It is well known that potential energy of fuel at 25°C on combustion is converted into the sensible heat pf products of combustion at the flame temperature. Products of combustion after transferring their heat to the furnace chamber exit the furnace. Heat carried by products of combustion depends on the temperature of the furnace; higher is the furnace temperature higher is the amount of heat carried by POC. It may range somewhere in between 40 to 60% of the calorific value of fuel. Heat of POC can be recovered either external to the furnace by installing a heat exchanger or internally by recirculating the POC into the flame in the furnace itself. The former is called external heat recovery and the later is internal heat recovery.

In the following we discuss the principles of external heat recovery of POC. Normally a heat exchanger is integrated with the furnace which captures and reuses the heat of POC simultaneously.

Thermodynamic Principles of Capture and re-use of Heat of POC

Capture and re-use of heat of POC must be integrated. A heat exchanger integrates capture and reuse of heat. In the heat exchanger hot fluid (POC) flows co-current or counter-current to cold fluid, say air. Both fluids are separated by a wall. Hot fluid enters the heat exchanger at temperature Th1 and exits at temperature T_{h2} ($T_{h2} < T_{h1}$). Wall is heated by the heat transferred from the hot POC. Cold fluid enters the heat exchanger at temperature T_{c1} and leaves at T_{c2} such that $T_{c2} > T_{c1}$.

Heat balance over an infinitesimally small element of length dx can be written at steady state

Heat lost by hot fluid = Heat gained by the cold fluid − Heat loss from the element to the surrounding

Let mh, and mc are mass of hot fluid and cold fluid, C_{Ph} and C_{Pc} are the specific heat of hot and cold fluid then we can write:

$$m_h \times C_{ph} \, xdT_h = m_c \times C_{pc} \, xdT_c - dQ$$

In eq. 2 dT_h and dT_c are the change in temperatures of hot and cold fluid at any position along the length of the exchanger.

In an ideal adiabatic-reversible heat exchange between hot and cold fluid, dQ = zero and the process is reversible when temperature difference between hot and cold fluid at any position along the length of the heat exchanger, i.e.$\Delta Ti = (dT_{hi} - dT_{ci}) = 0$ provided

$$m_h \times C_{Ph} = mc \times C_{Pc}.$$

This is possible when both fluids have infinite contact time, and separating wall has zero thermal resistance. In this situation the temperature difference between hot and cold fluid at any position will be very small and constant along the length of the heat exchanger.

Finite thermal resistance of the separating wall and flow rates of both fluids make the heat exchange irreversible. Finite flow rates of both fluids will have finite residence time depending on flow rates and as a result all the heat is not transferred from hot to cold fluid. Similarly finite thermal resistance of the wall will also limit the transfer of heat to the cold fluid. In such a situation for an adiabatic process.

$\Delta T_i = (dT_{hi} - dT_{ci})$ will be non-zero, but will have constant value when $m_h \times C_{Ph} = mc \times C_{Pc}$.

The practical result of the irreversibility is that the heat exchange is not complete and there is always some heat which is left with the POC on leaving the heat exchanger.

Difference in heat capacities of fluid will influence the heat exchange process. For example if $C_{Ph} > C_{Pc}$, cold fluid can be heated nearly to the entering temperature of hot fluid provided $m_h = m_c$.

Efficiency of Heat Exchangers

Thermodynamically thermal resistance of the wall, heat leaving the exchanger with POC influences the thermal efficiency of the heat exchanger.

$$\text{Overall thermal efficiency} = \frac{100 \times \text{Sensible heat in preheated air}}{\text{Sensible heat in flue gas}}$$

According to the definition of overall thermal efficiency, it appears that the air can be preheated to the temperature above the flue gas temperature since no upper limit is assigned to the temperature of the preheated air temperature. Thermodynamically, in heat exchange between hot flue gas and cold air, air can not be preheated to the temperature above the flue gas temperature.

$$\text{Efficiency limit} = \frac{100 \times \text{Sensible heat in air at hot flue gas temperature}}{\text{Sensible heat in hot flue gas}}$$

$$\text{Relative efficiency} = \frac{100 \times \text{Overall thermal efficiency}}{\text{efficiency limit}}$$

$$\text{Relative efficiency} = \frac{100 \times \text{Sensible heat in preheated air}}{\text{Sensible heat in air at hot flue gas temperature}}$$

Consider a heat exchanger which receives hot flue gas at 1600K and cold air at 298K. The hot flue gases leave the exchanger at 900K and cold air at 1373K. About 15% of the heat in flue gases is lost to the surroundings. The ratio of specific heat of fuel gas to air is 1.2.

Heat balance gives $\dfrac{m_c C_{pc}}{m_h C_{ph}} = 0.55$

Overall thermal efficiency by equation $\text{Relative efficiency} = \frac{100 \times \text{Overall thermal efficiency}}{\text{efficiency limit}}$ is 45.4%

Efficiency limit by above equation is 55%.

Example

Regenerator receives hot flue gases at $1400°C$ and cold air at $25°C$, the flue gases leave at $750°C$ and the air is preheated to $1100°C$. As estimated 15% of the heat given up by the flue gases is heat lost to the regenerator surroundings, and the rest (85%) is recovered in the preheated air. It may be assumed for estimating purposes that $C_p = 0.3$ for flue gases and $C_p = 0.25$ for air, independent of temperature. Estimate over all thermal efficiency, efficiency limit, and relative efficiency for this heat exchange operation.

Suppose now that the depth of the regenerator is increased to 2.5 times in such a way to double the heat exchange area while keeping constant the over-all heat transfer coefficient $U(\frac{B\,tu}{hr\,ft^{2\circ}F})$. The quantities and entering temperatures of the flue gases and air will be kept the same. Heat losses are same as that in a). Estimate for the enlarged regenerator (a) air preheat temperature, (b) over-all thermal efficiency and relative thermal efficiency

Solution:

(a) Heat balance: reference temperature 25°C

$$m_a\, C_{pa}(1100 - 25) = 0.85 m_f\, C_{pf}(1400 - 750)$$

$$\frac{m_a C_{pa}}{m_f C_{pf}} = 0.514$$

Overall thermal efficiency $= 40.18\%$.

Efficiency limit $= 51.4\%$.

Relative efficiency $= 79.4\%$.

(b) Air preheat temperature and exit temperature of flue gas are not known. Since quantities and entering temperatures of flue gas and air are same. We can write

$$\ln\left[\frac{T_{h_2}-25}{1400-T_{c_2}}\right] = 2.5 \times \ln\left[\frac{750-25}{1400-1100}\right]$$

$$T_{h_2} - 25 = 12698 - 9.07 T_{c_2}$$

Heat balance for the enlarged regenerator:

$$m_a\, C_{pa}(T_{c2}-25) = 0.85 m_f\, C_{pf}(1400-T_{h2})$$

In above equation, T_{c_2} and T_{h_2} are air and flue gas temperature at the exit of the regenerator.

Or $0.605\, T_{c2} - 15.11 = 1400 - T_{h2}$

By solving above equations,

We get $T_{c2} = 1335.8°C$ and $T_{h2} = 557°C$

Overall thermal efficiency $= 49\%$.

Relative efficiency $= 96\%$.

Sustainable use of Energy

Energy balance is a very powerful tool to address the issues of saving natural resources.

What can be done More Beyond Fuel Savings? Let us think.

Facts about use of natural energy resources

Fossil fuel based energy availability is associated with:

a) Discharge of CO_2 in the environment and other harmful gases like SO_2, NO_x etc. 1 kg mole of carbon discharges 1 kg mole of CO_2 in the environment. In other words 1 kg carbon produces 3.7 kg CO_2.

b) Large amount of heat is carried away by the products of combustion. In metal extraction processes at high temperatures, products like liquid metal, slag, gases and coke carry a large fraction of heat. Sensible heat is the issue of concern.In this connection it is important to understand quality of heat. Quality of heat is decided by the temperature. Higher is the temperature of discharged product, higher is the quality of heat.

Sensible Heat in Products

In thermo –mechanical processing, steel is heated to $1200^\circ C$. Steel is rolled at this temperature and then cooled by water. Large amounts of emissions are discharged.

In coke making coke is discharged at around $1200^\circ C$ and water cooled. Large amount of sensible heat is lost and as well as emission arising from water contaminants are discharged into atmosphere. Environmental pollution is a concern.

Sensible heat in waste gas may not be possible to use since waste gases contain fine solid particles. However potential energy can be utilized.

Energy Balance and Environment Cleanliness

Let me illustrate how energy balance can lead to environment cleanliness. Consider coke making technology.

Coke is an important source of chemical and thermal energy in integrated steel plants producing steel from iron ore. In fact 1 ton of steel would require around 940 kg coal (Assuming coke consumption is 500 kg/ton of hot metal), thus 10,000 tons of steel would require 6600 tons of coke (assuming 0.75 ton of steel is produced from 1ton hot metal). For annual production of few million tons of steel, Coke requirement would be very high. Let us consider coke-coke conversion materials and heat balance.

Consider heat balance of a by-product coke oven. Heat balance is given for 1000kg coal.

Heat input 8184×10^3 kcal

Calorific value of coke oven gas 704×10^3

(It is assumed that 40% of coke oven gas of calorific value $(113 \times 10^3$ kg cal / kg.mole) is used to produce heat in coke oven)

Heat out put	kcal
Calorific value of coke	5416x10³
Calorific value of tar	351x10³

Calorific value of coke oven gas	1053x10³
Sensible heat in coke	305x10³
Sensible heat in coke oven gas	174x10³

Analysis of Heat Output

Calorific values of coke, tar and coke oven gas can be utilized by combustion. Sensible heat in coke is notutilized since coke is wet quenched on wet quenching all sensible heat 305×10^3 kcal / 725kg coke is lost. In addition to this loss, air born coked us emissions 50kg/ton coke are produced. Also huge water is required say around $0.5 - 0.6\text{m}^3$ / ton coke. The contaminants in water are discharged in environment.

Energy Balance and Technology Development

The above analysis puts a pressure to develop a new technology which can capture and reuse the sensible heat in coke. This new technology will also contribute simultaneously to the cleanliness of environment.

Dry quenching technology (DQT) has been developed. The advantages are:

- Elimination of emissions
- Capturing heat and reuse- cogeneration.

DQT has to components

- Capture of heat in gas flowing counter current to coke
- Production of superheated steam in a boiler which can be used for example to run a turbine. Figure is the arrangement of various reactors in DQT.

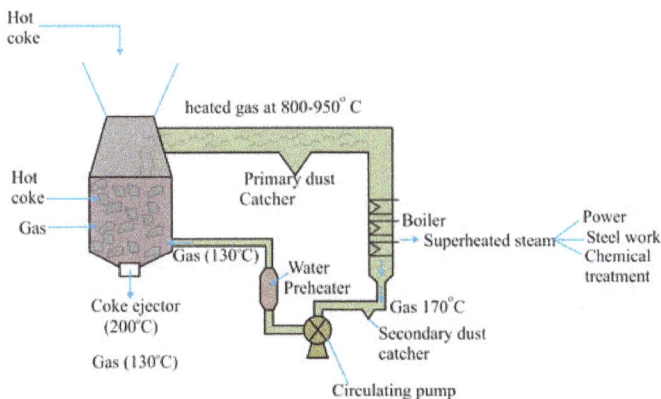

Arrangement of vessel to capture and reuse head in DQT.

References

- Pigott, Vincent C. (1999). The Archaeometallurgy of the Asian Old World. Philadelphia: University of Pennsylvania Museum of Archaeology and Anthropology. ISBN 0-924171-34-0, p. 191

- Wagner, Donald B. (2001). The State and the Iron Industry in Han China. Copenhagen: Nordic Institute of Asian Studies Publishing. ISBN 87-87062-83-6, pp. 75–76

- Jenkins, Barrie; Mullinger, Peter (2011-08-30). Industrial and Process Furnaces: Principles, Design and Operation. Butterworth-Heinemann. ISBN 9780080558066

- Gray, W.A.; Muller, R (1974). Engineering calculations in radiative heat transfer (1st ed.). Pergamon Press Ltd. ISBN 0-08-017786-7

- Fiveland, W.A., Crosbie, A.L., Smith A.M. and Smith, T.F. (Editors) (1991). Fundamentals of radiation heat transfer. American Society of Mechanical Engineers. ISBN 0-7918-0729-0

- Warring, R. H (1982). Handbook of valves, piping and pipelines (1st ed.). Gulf Publishing Company. ISBN 0-87201-885-7

- Whitehouse, R.C. (Editor) (1993). The valve and actuator user's manual. Mechanical Engineering Publications. ISBN 0-85298-805-2

- ASHRAE (1992). ASHRAE Handbook. Heating, ventilating and air-conditioning systems and equipment. ASHRAE. ISBN 0-910110-80-8. ISSN 1078-6066

- Perry, R.H. and Green, D.W. (Editors) (1997). Perry's Chemical Engineers' Handbook (7th ed.). McGraw-Hill. ISBN 0-07-049841-5

- Lieberman, P.; Lieberman, Elizabeth T (2003). Working Guide to Process Equipment (2nd ed.). McGraw-Hill. ISBN 0-07-139087-1

Permissions

Index